Traffic Calming in Practice

CONTENTS

PREFACE

Reducing vehicle speeds and securing environmental improvements, particularly within urban residential areas and villages, would probably be agreed by most communities as the highest priority for action by their local Highway Authority. Unfortunately in many cases this is where the agreement stops. The variety of individual aspirations, the range of techniques and materials, financial constraints and regulations, provide fertile ground for disagreement, controversy and delay. More positively, however, they also provide opportunities for genuine community involvement and partnership.

This book, drawing together practical experience from throughout the United Kingdom, is intended to broaden experience and understanding of traffic calming practice and to ensure that we continually improve the quality of both schemes and the procedures for their introduction.

I commend it to you.

Mike Kendrick
President of the County Surveyors Society

The Joint Working Group

Alan Lovell (Chairman)	*West Sussex County Council (CSS)*
Malcolm Baker	*Devon County Council*
Jim Bennett	*Northamptonshire County Council (CSS)*
Malcolm Bulpitt	*Kent County Council*
John Clayton	*Dover District Council (ACTO)*
Roy Endersby	*London Borough of Sutton (ALBES)*
John Martin	*Royal Borough of Kingston upon Thames*
Paul Primmer	*Wakefield Metropolitan District Council (AMDE)*
Mike Talbot	*Department of Transport*
Colin Thame (deceased)	*Department of Transport*
Eric Hinkley (Technical Editor)	*Highways & Transportation Consultant*

Consultees

The Working Party thank the following organisations who contributed to the book through their comments on the pre-publication draft.

Association of Chief Police Officers of England, Wales and Northern Ireland
Association of Transport Coordinating Officers
Chief and Assistant Chief Fire Officers Association
Confederation of Passenger Transport UK
Department of Transport
Home Office (Fire Brigades)
London Transport Buses
NHS Executive, Department of Health (Ambulance Service)

TRAFFIC CALMING
IN PRACTICE

County Surveyors Society
Department of Transport
Association of London Borough Engineers and Surveyors
Association of Metropolitan District Engineers
Association of Chief Technical Officers

Published by Landor Publishing Ltd.
Quadrant House
250 Kennington Lane
London SE11 5RD

Designed and produced by
Kennington Publishing Services
London

First published November 1994

ISBN: 1 899650 00 8

INTRODUCTION

The words 'traffic calming' are used a lot nowadays and it is apparent that many of the people who use them have their own view of what they mean. This can create problems for those engaged in this field of work, particularly when the public's expectation of what traffic calming can achieve is over optimistic. Nevertheless 'traffic calming' has now entered into popular parlance and most highway authorities are responding to requests to introduce it to tackle a variety of problems, be they real or perceived. In a sense, perceived problems are as real as those that can be quantified because they are important to the people who are experiencig the intrusion that motor vehicles make on their daily lives.

For some people the words 'road humps' have become synonymous with 'traffic calming' but this is a misconception. A wide variety of techniques in addition to humps have now been successfully applied to reduce traffic speeds and the risk of accidents, as well as acting as a deterrent to traffic choosing a particular route.

This book has been produced in response to requests from practitioners for practical advice on how to go about traffic calming; and for information on what techniques have proved successful, or indeed unsuccessful, elsewhere. The main objective, therefore, has been to gather together a large number of examples of schemes from all parts of the country. There are profiles of more than 80 schemes in town centre, residential and rural locations selected to show the variety of techniques available. For each case study a simple format has been chosen to give an overview of the local 'problem', the techniques applied and the degree of success achieved. We have also included a contact name and telephone number so that further details can be obtained and information exchanged.

The opening chapters not only introduce traffic calming but go into detail on objectives, legislation, consultations and assessment of priorities. There is advice on how to go about developing and implementing a successful traffic calming scheme and the importance of 'before' and 'after' studies. A commentary on the case studies is included together with quick reference tables summarising their effectiveness, cost and main features. This should be particularly useful to those people tackling these matters for the first time and for those looking for new techniques.

If there is one piece of advice I would give it is to stress the importance of establishing a dialogue with local people and the other bodies involved, at the earliest possible time; and to continue this through design, implementation and monitoring. Proper dialogue means being prepared to listen, to change both your mind and the scheme. It also means ensuring most people understand the proposals. It is not unknown for local people to be highly critical of the scheme after construction because it is not what they thought they were getting when it was presented to them. Although traffic calming measures can be cheap and effective, the reverse can also be true and altering a scheme afterwards can be very costly.

This book is the result of contributions from a large number of people and I should like to thank my colleagues from the County Surveyors' Society, Association of London Borough Engineers & Surveyors, Association of Metropolitan District Engineers and Association of Chief Technical Officers for providing the case studies. Special thanks are due to Alastair Jefford and Malcolm Bulpitt of Kent County Council, Mike Talbot and the late Colin Thame of the Department of Transport, John Martin of the Royal Borough of Kingston upon Thames and Paul Primmer of Wakefield Metropolitan District Council for their input and support throughout the project.

Finally, I should say that traffic calming is very much a developing art. If you have particular examples or experiences you would like to submit for a possible second edition of this book then do let me know.

Alan Lovell
Chairman of the Joint Working Group
County Surveyor to West Sussex County Council

CHAPTER 1

The Framework for Traffic Calming in the UK

Traffic calming is the term that has come to be used in the UK for the application of traffic engineering and other physical measures designed to control traffic speeds and encourage driving behaviour appropriate to the environment. It is important to remember that the measures can include the use of 'traditional' traffic engineering techniques, as well as a variety of other measures more recently introduced into the UK such as road humps, gateways, speed cameras and various forms of horizontal deflections. Also included are the use of visual messages, either explicit such as signing, or implicit such as landscaping.

The essence of traffic calming lies not in the use of specific measures but in the overall objectives to create safer roads and better environmental conditions. The strategic objectives are:

- improving driver behaviour, concentration and awareness,
- reducing speed, disturbance and anxiety,
- enhancing the environment.

Influencing human behaviour, the main contributory factor to road traffic accidents, can be achieved directly by education, training and publicity as well as indirectly, by using environmental and engineering measures. The environmental and engineering measures aim to create situations which induce safe road user behaviour and reduce the scope for unsafe conditions to occur. Traffic calming, using both environmental and engineering measures, has an important role to play in influencing behaviour, especially that of inconsiderate drivers who drive too fast for prevailing conditions or with lack of care and attention.

For many years engineers have introduced traffic management schemes using conventional techniques such as banned turns, road closures and one-way streets to reduce through traffic in particular areas; these remain valid measures. The more recent rise of interest in traffic calming can be traced back to the introduction of road humps. The restrictive nature of the initial regulations in the UK generally limited road humps to residential areas and such areas were also seen as a priority by local authorities.

There have since been significant developments in the variety of vertical deflection devices and in the range of other measures available to traffic engineers which can be of value in all locations and the Traffic Calming Act 1992, and subsequent regulations, give a legal framework for:

- footway buildouts
- gateways
- rumble devices
- islands
- overrun areas
- chicanes
- pinch points

The range of options is continuing to increase as

The majority of traffic calming schemes in the UK have been applied in urban residential areas, often aimed at reducing child pedestrian casualties. Picture: Crumpsall Green, Manchester (Case Study 42).

local authorities develop their own ideas experimentally and look to experience elsewhere, particularly in continental Europe.

Traffic calming can be applied to many different situations although the greatest focus is in urban residential areas. Over half of the studies featured in this book are from such sites. It is also being applied on distributor roads and in the centre of historic towns as well as in towns and villages which have been bypassed. Main roads through the centre of small villages and roads in environmentally sensitive locations such as national parks may also be suitable for traffic calming. It is important to acknowledge that no 'off-the-shelf' solutions are available. A fundamental tenet of traffic calming is that any scheme should be designed to fit the particular circumstances of its location.

As the public has become more aware of traffic calming, and as the legislative basis and technical experience have developed, calls for traffic calming in one form or another have burgeoned. Most schemes are intended to tackle problems over a street or an area rather than at a single point for which the application of Accident Investigation and Prevention measures may be better suited.

Traffic calming cannot provide a solution to all traffic problems and typically in any area there are far more demands for schemes than can be implemented within current resource levels. Authorities are therefore faced with the difficult and often controversial task of allocating priorities and this needs to be done using a clear set of objectives.

Traffic Calming Objectives

Safety is a key objective for virtually all traffic calming schemes, not only in terms of accidents recorded, but also because of the degree of danger felt by people using the streets concerned.

The initial interest in traffic calming stemmed from a number of considerations. Prominent was safety with increasing concern being expressed at the numbers of road accidents, particularly among the more vulnerable members of society. Another worry was rising traffic volumes causing vehicles to rat run through areas not suited to through traffic. Also there were concerns about the widespread disregard for speed limits. All of these factors have contributed to extensive public pressures for measures to civilize traffic, particularly in sensitive areas.

Environmental improvement has become an objective in its own right for many traffic calming schemes although such an approach will not be possible or, indeed, affordable everywhere.

Concern over the adverse environmental and health effects of traffic has increased considerably in recent years. Reducing the speed and volume of traffic can contribute to a better local environment, but it may often be possible, through imaginative design and the use of appropriate materials, to provide further enhancement.

The development of traffic calming measures into a holistic approach to the use of street space can change the character of a street which, in turn, helps to encourage appropriate driving behaviour. Traffic calming is thus increasingly incorporating wider quality of life issues which, whilst inherently difficult to quantify, are regarded as important by many people.

Value for money remains an important objective, underlining the importance of effective design and sound assessment when dealing with the high demand for traffic calming schemes.

Traffic calming schemes are, generally, no less expensive than traditional traffic management schemes. Some may be cheaper and still have an effect but because of the need to give greater consideration to the materials being used, and their maintenance, they may often cost more. Costs can vary considerably depending on the features being employed, and the materials used. The case studies in this book have costs varying from under £3,000 for a single measure as in Tuly Street, Barnstaple in Devon (Case Study 6) to several over £100,000 per kilometre of road treated, such as Carfax, Horsham in West Sussex (Case Study 1), High Street, Buntingford in Hertfordshire (Case Study 22), and Nunsthorpe Estate, Grimsby in South Humberside (Case Study 66).

Policy Framework

Government policy encourages highway authorities to develop traffic calming schemes, particularly in the context of its objective to reduce road casualties by a third by the year 2000. It also encourages a view of traffic calming that encompasses the environmental as well as the safety role; and the use of alternative modes of transport to the car.

Most local authorities also view traffic calming as an important element in their transport strategies and plans and increasing sums of money are being allocated for it. This fits in well with the relatively new 'package approach' to transport and its funding which has been developed by the Department of Transport (DOT). Local authorities are invited to put forward strategies for their areas with associated spending plans and traffic calming should be considered an integral part of such strategies and plans.

It is preferable that traffic calming schemes are developed in the context of a strategy for the road network as a whole, and the package approach should help schemes to be planned on an area-wide basis as part of this broad perspective.

The Urban Safety Management approach[52] provides a sound basis for developing a such a strategy within urban areas. This involves establishing a hierarchy within the network, identifying the occurrence of accidents and perceptions of safety in each section, and hence developing a Safety Strategy. This approach allows the role of each road to be defined and appropriate traffic calming schemes tailored to that role.

Legislative Framework

Local Highway Authorities in England and Wales are given a general power to improve highways under Section 62 of the Highways Act 1980. Section 75 of the Act empowers a highway authority to vary the relative widths of carriageways and any footway and Section 77 permits the level of a road to be raised, lowered or otherwise altered. However this general power cannot be relied upon when constructing the sort of localised and relatively sudden vertical deflection which a road hump produces.

The current legal basis for road humps, which are still the most commonly used traffic calming measures, lies in sections 90A to 90F of the Highways Act 1980 (as amended by the Road Traffic Act 1991) together with the Highways (Road Humps) Regulations 1990 which provided for considerable relaxation from the earlier 1983 and 1986 Regulations. The new regulations allow for the construction of flat topped humps as well as round topped humps within 30 mph speed limit areas, effectively permitting the use of raised junctions.

Legislation to remove previous anomalies and resulting uncertainty about the use of different features for traffic calming purposes was enacted in the Traffic Calming Act 1992 which inserts additional sections 90G, 90H and 90I into the Highways Act 1980. The Highways (Traffic Calming) Regulation 1993 allows authorities to implement a range of traffic calming measures for environmental as well as safety reasons. The changes to the Highways Act also empower the Secretary of State to specially authorise features where existing powers are not considered sufficient.

The procedures for obtaining consent for 20 mph zones are set out in Circular Roads 4/90.(8) Where consent for a 20 mph zone has been granted or where special authorisation for non-conforming road humps is obtained certain requirements are further relaxed, for example location, signing and lighting.

Road humps have been the most popular traffic calming measure in the UK, used here at Sittingbourne in Kent (Case Study 68).

The devices to be used in the zones are for the authority to decide. An important consideration is that they must be self enforcing in bringing average speeds down to or below 20 mph. Furthermore no road in the 20 mph zone may be more than 1 km from the boundary of the zone.

Where the speed limit exceeds 30 mph, special authorisation for vertical deflections may also be sought from the DOT. However strong justification, including a safety evaluation of the need for such measures, is required.

Traffic signs and road markings required in association with road humps or other measures, and those which could be used as measures in their own right, are covered in The Traffic Signs Regulations and General Directions 1994. There is currently no statutory prescribed sign to indicate a traffic calmed area but there is a prescribed 20 mph speed limit zone sign.

Advice on the use of the various traffic calming measures and the procedures to be followed is also given in a number of Traffic Advisory Leaflets published by the Department of Transport which are listed in the Bibliography at the end of the book.

Notwithstanding the powers granted to highway authorities by the above legislation, it is important that authorities satisfy themselves that any measure used is safe and legally permissible in relation to their fundamental duty to maintain the safe and free flow of traffic.

The Need for Consultation

Traffic calming schemes should be designed to fit in with the surrounding environment. What is appropriate for a residential area may not be appropriate for an historic town centre and vice versa. Scheme development may therefore involve traffic engineers, town planners, landscape architects and other disciplines, especially when environmental objectives are important.

Designers must take into account not only those who use the road but also those who live and work near it. Among road users particular attention should be given to the needs of emergency services and bus operators who may have difficulties in negotiating some traffic calming measures. Early consultations with these organisations are essential. Consultation with the police is a statutory requirement for road humps, as it is for traffic calming measures implemented under the Traffic Calming Regulations.

Other problems may affect specific groups in certain locations. For instance, near churches and crematoriums where long-wheel-base hearses can ground on humps unless a minimum plateau length of 4m is provided (as found in Case Study 47, Fenton Area, Stoke on Trent). The interests of vulnerable road users, cyclists[44], pedestrians and motorcyclists require particular attention, and the needs of people with disabilities must also be taken into account[51]. In the vicinity of schools and shops consultation with headteachers, shopkeepers and appropriate organisations will be necessary.

Frontagers will be concerned about access to premises and parking and loading as well as the environmental impact of a scheme. The extent of consultation with these groups will vary according to the circumstances, but early consultation is necessary to develop a scheme that can achieve wide acceptance. Even low cost measures can be costly to remove, as is illustrated by Christchurch Road, Bexley in London (Case Study 60). Follow up consultation is desirable during the monitoring period

and time and resources need to be allowed for this within authorities' programmes.

The effects of traffic calming measures must also be considered over a wide area as measures in one location can cause traffic to divert to other roads, and this should be reflected in the area adopted for consultations. More advice on consultation is given in Chapter 2.

Funding

The potential sources of funding for traffic calming schemes are the local authority's own resources, directly or through its Transport Policies and Programmes (TPP) submissions to the DOT, from the private sector and occasionally from residents. Within the TPP system, schemes under £2m (which effectively covers all traffic calming schemes) are now funded through capital credit allocations rather than Transport Supplementary Grant (TSG). However some traffic calming schemes may qualify for 'earmarked TSG allocations' as local safety schemes depending on the estimated level of accident reduction. The DOT's 'package approach' to transport planning and funding has already been mentioned. This provides for funding to be decided on the package as a whole, with the allocation of priorities among schemes within the package a matter for the local authority.

Opportunities to obtain contributions from the private sector towards traffic calming schemes, particularly in town or local shopping centres, may also arise. Examples, both large and small, are Carfax, Horsham, West Sussex (Case Study 1); Shenley Road, Borehamwood, Hertfordshire (Case Study 3), and Tuly Street, Barnstaple, Devon (Case Study 6). The most promising opportunities are likely to arise in association with development, so traffic calming objectives and criteria should be included in development briefs. For redevelopment or regeneration projects in inner city areas some funding may be possible through various Department of the Environment (DOE) urban programmes as part of a comprehensive redevelopment package.

Traffic calming can also be built into areas being developed for residential uses as recommended in Design Bulletin 32.[45] In Strathclyde, for example, a future requirement for new residential roads will be that they conform to a traffic calmed

Public acceptance of a traffic calming scheme is essential for effectiveness. Environmental improvement features can be helpful in making a scheme popular with residents and local businesses. Picture: Buntingford, Hertfordshire (Case Study 22).

design which will allow the area to be considered for 20 mph zone status. Wiltshire also has traffic calming guidelines for new residential areas, as have other authorities in their Local Design Guides.

Incorporating traffic calming measures in new schemes or redevelopment may add a little to the initial cost but will be considerably cheaper than having to add them later. Similarly if the implementation of traffic calming measures can be co-ordinated with maintenance work, as in York, savings may be possible.

Occasionally, proposals for privately funded traffic calming schemes are put forward by residents. In considering such requests local authorities should ensure that the full costs are taken into account, because such schemes will divert staff from predetermined and approved priorities. Costs considered should include the not insignificant design, legal, supervision and consultation costs as well as attributable increases in future maintenance costs. These last may be funded by way of a commuted sum payment.

Assessment and Priorities

Local authorities are now facing increasing demands for traffic calming schemes generated by both their own studies in pursuit of their safety and environmental objectives, and by requests from all sections of the community who, sometimes unrealistically, see traffic calming as the solution to the problems of their locality. Faced with this situation, it is important that authorities adopt a systematic approach to assessing options and assigning priorities in order to make best use of limited resources.

There is rarely a single traffic calming 'solution' to any situation. There are choices to be made between types of measures, materials to be used, area to be covered, level of environmental enhancement etc, all with cost and other implications. The use of a structured assessment framework should assist all those involved in developing schemes to compare options against agreed objectives. The framework may also have a role during consultation with the public as well as for the professionals involved.

Any assessment framework needs to recognise that traffic calming schemes in different types of locations may have different local objectives or a different balance between objectives. This may mean that assessment methods will vary for different types of schemes; or, if the methods are the same, the relative weightings given to different factors may vary. In either case it is unlikely that a scheme in, say, a historic town centre can be readily compared with a scheme in a village on the main road network.

Schemes will need to be categorised. Priorities within categories may be derived by a common assessment method but choices between categories are likely to be more subjective. All methods will involve a significant degree of judgement given the difficulty of quantifying some of the relevant factors.

Not only is there no single solution to a problem there is no single approach to assessment that will be applicable in all situations. It is for each authority to decide what is appropriate for its own local circumstances. It will be important, however, to ensure that whatever method is followed it is capable of being defended against calls for changes in the priorities. The use of a systematic approach that has been endorsed by elected members can be particularly useful in such circumstances.

Safety is, in most cases, the primary objective of traffic calming schemes. Within urban areas, the Urban Safety Management approach[52] to the road network provides a sound framework within which to assess individual traffic calming schemes. It establishes the role of each part of the network, allowing schemes to be categorised and appropriate objectives set for each category. It also provides the essential data on accidents and risk that can help to give a measure of the problem and the potential benefits.

It is important to keep in mind that there are likely to be some disbenefits as well as benefits associated with each scheme so this must be reflected in the way factors are considered.

Assessment Framework

A number of authorities have developed methods for assessing traffic calming schemes in their own areas. These range from those based solely on the accident record through to more complex methods which take account of physical characteristics of the road and environmental factors. Systems are usually based on a range of factors drawn from those set out in the 'Traffic calming scheme assessment framework' panel overleaf. Authorities must decide which factors and weightings are relevant to the decisions they are making at any one time.

The precise way in which the factors are used in the assessment will vary with different assessment methods but it is vital that two important aspects are clearly distinguished and analysed. First there is the absolute value of the factor, which generally gives an indication of the size of the problem. Second there is the change that the proposed scheme is expected to produce, which is a measure of the benefit (or disbenefit). The two will often point the same way, but not always. A particular scheme may have little effect even though the site has a serious problem.

In seeking value for money, making realistic estimates of the benefits to be achieved is an important part of the assessment process. That process should also take account of prospective

Traffic calming scheme assessment framework

Objectives
- achievement of Road Safety Plan objectives
- achievement of Transport Policy and Strategy objectives
- achievement of Planning and Environment objectives

Safety
- total number of accidents
- number of accidents to vulnerable road users
- number of accidents involving speed
- perceived risk and exposure to traffic (particularly by vulnerable groups)

Traffic characteristics
- day time and night time vehicle flow
- volume of through traffic
- number of HGVs
- traffic speeds
- pedestrian flows
- cycle flows
- public transport routes
- emergency vehicle routes

Physical characteristics
- width of road
- alignment (gradient, radii, etc)
- footway provision
- provision for cyclists and pedestrians
- provision for people with disabilities
- parking provision
- location of bus stops
- sensitive frontage activities (e.g. schools, hospitals, shops)

Environment
- traffic noise and ambient noise
- severance of the community by the road
- severance of cycle routes and networks
- number of residents affected
- distance of houses from road
- other properties affected

Value for Money
- costs of construction and maintenance
- cost of design and consultation (before and after)
- benefits (taking account of diversion to other routes - time and distance)
- First Year Rate of Return or Net Present Value
- cost of alternative solutions

changes under the do-minimum situation, e.g. as a result of a new bypass. The assessment should explicitly identify any disbenefits of a scheme to particular groups or locations even where, overall, there may be a net benefit. This will be particularly important during the development of options and at the consultation stages.

Allocating priorities

In some assessment methods absolute values and/or changes in these factors are assigned numerical scores based on predetermined criteria or informed judgement. The scores will then be multiplied by an agreed set of weightings to provide an overall score for each option or scheme. This can be used for ranking the benefits obtained from different schemes. Given the safety objectives of traffic calming, a high weighting will usually be given to accident reduction. It may also be appropriate to give weight to schemes which improve accessibility conditions for vulnerable road users over and above simply reducing accident figures. Some fine tuning of the weightings used may be appropriate over time to reflect the evolution of local concerns and monitored experience. In any assessment framework the factors, criteria and weightings to be used must be clearly defined and, preferably, agreed by elected members.

The need for consultation when developing traffic calming schemes frequently leads to an iterative process of design, which is time and resource consuming at the design stage. In order to concentrate the design effort most efficiently it may be appropriate to have a multi stage assessment method for determining priority areas. The first stage would be an assessment based on a limited number of factors, principally safety related; indeed it may be appropriate to sift the potential schemes initially solely on the basis of accident and speed reduction potential. The next stage would be to look in more detail at a wider range of factors for those schemes which pass the first sift.

Any assessment method should endeavour to put competing projects on an equivalent basis so that larger and smaller projects can be fairly compared. This may mean incorporating comparative measures such as the length of road or number of houses affected by different schemes in order to obtain a standard measure of cost and benefit.

Authorities should recognise that it is unlikely that an assessment structure can be devised to equitably handle all schemes, especially both urban and rural schemes. *An assessment based on urban oriented criteria will leave rural schemes forever at the bottom of the priority list.* The objectives, and hence the assessment criteria, are likely to be significantly different for urban and rural areas and authorities may, therefore, find it necessary to establish separate programmes and budgets for urban and rural schemes.

It is probable that even within similar area categories the separation of safety-led schemes (usually the majority) and environmentally-led schemes, will be useful, if only in terms of potential funding through the TPP system. However, the package approach to funding would not preclude additional monies being devoted to safety-oriented traffic calming schemes over and above the amounts earmarked specifically for local safety schemes.

Ultimately traffic calming is very much a local issue and it is for the local authority to decide on its priorities. However, *it is strongly recommended that authorities establish some form of systematic assessment of traffic calming schemes to help ensure that resources are targeted to best effect.*

Monitoring

'Before' and 'after' monitoring of schemes is also an important discipline. Unfortunately it is often overlooked or given low priority. It is only by systematic monitoring that an authority can check whether its money has been spent effectively and can build up knowledge of the effectiveness of different measures to assist future decisions. This is particularly important in the current situation when experience of some traffic calming measures in the UK is still limited.

Sharing the results of monitoring will help guide other authorities considering similar measures and enable greater confidence to be built up about the role and performance of each type of measure. Therefore it is strongly recommended that all schemes be carefully monitored. For innovative measures not explicitly permitted under the Highways (Road Humps) Regulations, the Highways (Traffic Calming) Regulations or the Traffic Signs Regulations and General Directions but which may nonetheless be given site specific authorisation by the Secretary of State, systematic monitoring may be required as a condition of such authorisation. Currently the DOT is discussing the possible deregulation of traffic calming to give local authorities more flexibility to try innovative measures. Such a move would, of course,

mean greater local responsibility and accountability.

Key parameters to monitor will always be accidents and traffic speeds although others, such as traffic flow and diversion to other routes, may also be important in particular circumstances. The factors monitored should reflect the objectives driving the scheme and those used in assessing priorities for funding. Public information on, and reaction to, actual effectiveness is also vitally important, as monitored results may diverge from the expected effects that were highlighted at the consultation stages. Establishing what the public's expectations and perceptions were, and to what extent they have or have not been realised, will be helpful for future schemes.

The extent of monitoring will depend on local circumstances and, subject to any requirements for special authorisations, will be for the authority to decide. Advice can be sought from TRL and the Driver Information and Traffic Management Division of the Department of Transport, but the following should be considered:

- the number and location of survey sites (e.g. to detect any diversion)

- the need for a control site (e.g. in relation to overall accident trends)

- spot speeds or average speeds through a scheme

- short term and long term effects (does the impact wear off over time?)

- number of measurements needed for statistical reliability.

The opportunity now exists for traffic calming to move from being a purely ameliorative measure to playing an integral part in the transport strategies and plans of all authorities. The new awareness, the wider objectives, the clearer legislative position, the results of research and the new funding approach have all helped to move traffic calming to the forefront of local transport policymaking. The following chapters suggest how to go about preparing schemes, and illustrate what can be achieved.

FIGURE 1. STAGES IN PROMOTING A TRAFFIC CALMING SCHEME

Establish the policy framework in Road Safety Plan, TPP, UDP, Structure Plan

Establish role of roads on road network and develop a Road Hierarchy. Designate roads not to be traffic calmed and determine appropriate types of measures for other main roads in consultation with emergency services and bus operators.

Identify need and priority for action. Is it environmental and/or accident reduction?

Prepare concept designs.

Informal consultation with emergency services and bus operators.

Initial consultation with local people - residents, businesses and community organisations.

Evaluate feedback. Do benefits outweigh disadvantages?

No   **Yes**

Commission detailed design including modifications indicated by consultation.

Carry out formal consultation with emergency services, bus operators, elected members, the public, Chambers of Commerce, road haulage industry, motoring organisations, commercial and other groups in the area.

No   Evaluate feedback
Do benefits outweigh disadvantages?

Yes

Obtain formal authority to implement including special authorisation (DOT) if needed.

Abandon scheme

Programme scheme.

Issue notification of implementation.

Monitor (including opinion gathering as well as factual data) and Review after implementation.

CHAPTER 2

Development and Implementation

This chapter gives advice on developing a scheme that is effective in achieving its objectives and also acceptable to most people and organisations likely to be affected. It is not so much about 'selling' a traffic calming scheme as listening to other points of view and responding constructively. This chapter does not deal with detailed design or the precise mechanics of scheme presentation because each authority will have its own tried and tested methods for these. Some of the topics discussed in Chapter 1 recur, quite deliberately, to give them added emphasis.

Because it is a very common and seemingly intractable problem, specific advice is given on ways to resolve the conflicts of interests that can occur between the objectives of traffic calming and those of the emergency services and bus operators.

Setting the Scene

A number of local authorities are actively involved in wide ranging consultations about strategic transport issues and an increasing number are conducting 'awareness' campaigns aimed at changing attitudes and travel behaviour. *A principal aim of these is to encourage responsibility in use of the car.* Such campaigns, especially those involving the community, provide an extremely helpful framework in advance of consultations on a specific traffic calming scheme, or indeed any other transport project or issue.

On a different level, 'partnership' projects are being used around the country to increase local understanding, define problems and develop options for action. 'Planning for Real' and 'Traffic Games' are being used by a few authorities as a way of including local people in the develop-ment of local traffic schemes. Some experimental schemes have included residents working with officers on assessment and evaluation.

Nationwide there are a multiplicity of joint steering groups, local task forces, area committees and the like bringing together professionals and local people. *The development of trust and credibility between professionals and their customers is important* so that discussion on specific projects takes place in a constructive atmosphere. All of this however is very time consuming and can be relatively expensive in comparison with the cost of the measures themselves. It is for each authority to decide the extent to which it can go with strategic and scheme consultations.

Scheme Objectives

It is one of the realities of life that someone or some organisation will be unhappy with any proposal. The original purposes of a scheme must be made clear so that, if necessary, even the objectives can be modified in the light of comments that arise during

Consultation with bus operators and emergency services is essential to ensure that traffic calming schemes take into account the needs of these essential road uses. Picture: Pan Estate, Newport, Isle of Wight (Case Study 67).

consultation. If changes are to be made to any aspect of a scheme the sooner this is done, with reasoned explanations, the better. As Chapter 1 explains traffic calming schemes are, in broad terms, promoted either to reduce speeds and personal injury accidents or for environmental purposes. Both are entirely valid and both may apply in some cases but, faced with opposition over measures in pursuit of these aims, the promoter needs to consider how to respond.

If a scheme is being proposed for environmental reasons and there is significant local opposition, particularly from people who should benefit, then the authority's continued promotion of it must be in serious doubt. The authority may decide to drop the scheme or at least rethink the proposal and make amendments.

If, however, the scheme is an accident remedial measure, then the highway authority would probably wish to continue with it, albeit perhaps with modifications, notwithstanding local opposition. For those schemes with combined objectives, promoters should be clear whether the main objective is amenity improvement or accident reduction.

Basic Implementation Principles

Although this book is not a guide to detailed design a few basic principles are worth mentioning even though some designers may feel these are statements of the obvious.

The safety or environmental problems being addressed, be they real or perceived, need to be analysed to establish likely causes. For example, are the recorded accidents speed related or are there more significant factors to be considered which may or may not be susceptible to traffic calming treatment? Analysis requires, as a minimum, robust 'before' data on traffic flows, accidents, speeds and observations on site by experienced staff. There will usually be more than one possible solution and each different location may require different environmental treatment as well as different measures and materials. Some options may offer a better chance of improving driver and other road user behaviour, an important and very difficult task.

Consideration should be given to using traffic engineering techniques either separately or in combination with traffic calming features such as road closures, banned turns, one-way streets and waiting restrictions. A co-operative multi-disciplinary analysis and design approach is desirable in most cases, especially in town centres and other environmentally sensitive locations. This may appear to lengthen the design process but should reduce the need for changes at a later stage. Regular liaison with the rele-

vant planning authority should be the norm and started early.

Attention to detail in design, workmanship and materials from the maintenance viewpoint as well as end product appearance is important. Care needs to be taken to ensure that the materials, construction and layout are appropriate to their use because a number of schemes have required modification shortly after coming into use. Such situations can be costly to rectify. Some examples of post opening problems which required subsequent modification were found with The Parade, Leamington Spa, Warwickshire (Case Study 11); West Bretton, West Yorkshire (Case Study 33); Selsey, West Sussex (Case Study 37); and Christchurch Road, Bexley, London (Case Study 60).

From time to time it is important to check total cost estimates, value for money and priority ratings, as these can change significantly during the development of a scheme.

Although the works themselves are often small in scale they can attract a great deal of public comment and criticism if they are not planned and undertaken with care. By its nature, construction work is often carried out in the midst of other day-to-day activities and can cause noise and inconvenience to passers-by, residents and shopkeepers. It is very important that contacts are made before work is started and that workable channels of communication with the community are set up and used. For example whilst work is in progress most authorities now erect display boards on site with contact telephone numbers. Others distribute letters in the area prior to the commencement of works and follow these up with progress reports. If the works are programmed to last several months some authorities set up monthly liaison meetings of local representatives, including the local news reporter, to discuss and publicise progress.

Working with the Emergency Services and Bus Operators

A widespread problem currently exists between professionals working on traffic calming and professionals involved with ambulances, fire appliances and buses. There are real difficulties being experienced by the emergency services and bus operators and, ideally, these need to be addressed well before consultations take place on individual traffic calming projects.

Emergency Services: For the emergency services the concerns are twofold. First the reduction in speed increases the time it takes for a fire appliance or ambulance to reach an incident, and in some areas they will already be near the limit of Home Office and Health Service call-out criteria.[34, 51] Second when ambulances need to take a patient to hospital road humps, usually the most effective of speed reducing measures, can cause difficulties for ambulance staff and their equipment as well as considerable discomfort for patients.

In some areas the emergency services, as well as objecting to humps, flat or round topped, also object to chicanes and other horizontal deflections, particularly where these are closely spaced. On the other hand there are some emergency services that accept vertical deflections on certain routes provided they are spaced no closer than 100m. However, in the majority of cases where traffic calming has to be introduced horizontal measures are preferred to road humps.

Speed cushions lend themselves to a variety of layouts, especially where bus and emergency service access is needed. Here, the cushion within an offset narrowing creates a mild chicane effect. Camp Hill Area, Nuneaton, Warwickshire (Case Study 45).

The Police: The police are generally supportive of traffic calming schemes but they too have target times to meet when attending an incident in order to fulfil Citizens' Charter obligations. Since the police reach the majority of incidents by car, traffic calming generally does not have the same impact on police service delivery as it does on the fire and ambulance services. Nonetheless it is essential to seek the advice and co-operation of the police at a very early stage in the development of a scheme, including their views on accidents, traffic movements and speeds. The police will be very concerned if any proposals require a significant increase in police traffic and parking enforcement.

All the emergency services appreciate that the main objective of traffic calming is to reduce the number and the severity of casualties, but they are concerned that the widespread introduction of such measures could undermine the quality of service they are required to provide and want to achieve. At present no data exists to enable a trade-off to be attempted between improvements in road safety and disbenefits to ambulances and fire appliances as a result of traffic calming.

Bus Services: Similar problems apply to bus services. The introduction of traffic calming can lengthen journey times such that economic timetables cannot be followed. A few operators have withdrawn or rerouted bus services after the introduction of traffic calming schemes.

Bus passengers find the quality of ride is worse on traffic calmed streets and can experience difficulties when standing or moving along the bus as it negotiates a road hump. Such difficulties can be acute for elderly and infirm passengers. Relatively speaking there tends to be a greater loss in riding quality for passengers on buses than for private car occupants and, of course, car drivers usually have the option of choosing an alternative uncalmed route.

Considerable discomfort can be caused for bus drivers, who may have to negotiate the humps several times each hour for several hours each day, and many operators believe that wear and tear on vehicles is also increased by frequent negotiation of road humps. The bus industry intends to gradually introduce more low floor buses and this is likely to increase the problems posed by road humps.

Whilst many authorities are continuing to use road humps on bus routes, an increasing number use them only as a last resort when all other methods of calming have been rejected.

Research into Humps and Cushions

Recent research into road humps for the DOT by the Transport Research Laboratory (TRL)[60] describes types of vertical deflections together with an assessment of their effectiveness. The most common type is 50mm to 100mm in height with a flat topped or circular profile. The study found that speeds over humps have increased since their introduction in the 1970s, so closer spacing may now be required to ensure speeds remain low.

The TRL also conducted tests, for the DOT, on design of humps and speed cushions.[63]. The latter do not extend the full width of the road, but should be narrow enough to fit within the wider wheelbases of buses and emergency vehicles, while preventing cars from passing over at speed. On-road trials of speed cushions are being carried out in York [50] and a further report was published in September 1994. The results indicate that speed cushions can be used effectively on bus and emergency routes with these operators preferring narrower cushions in the region of 1.6m width.

A variety of speed cushion layouts are possible to suit local circumstances. Cushions can be used singly, in pairs or threes, and can be combined with refuges and protected parking arrangements. Careful design is necessary where there is parking, as parked vehicles can prevent buses being able to straddle a cushion as intended, and the profile then encountered may be more severe than a road hump. London Transport Buses has produced an update on its guidelines on the impact of traffic calming on buses[72] which outlines a number of recommended layouts. Although midi and mini buses cannot straddle the wider cushions of 1.8-1.9 m, their use may be appropriate where the main problem is excessive car speeds and if the frequency of bus and emergency service use is low.

A Road Hierarchy Approach to Traffic Calming

Because of the problems caused to emergency and bus services and complaints from local residents, at least seven authorities are known to be seeking ways of calming traffic without the use of road humps and other vertical deflections. Some do not use vertical deflections on main and distributor roads nor on major access roads used by bus services except, like Kent, as a last resort. The approach outlined below makes use of the conventional Road Hierarchy and is of benefit when considering traffic management schemes as well as traffic calming.

In West Sussex progress is being made with the emergency services and bus operators by taking the Road Hierarchy principle[52] one step further. After consultation with the emergency and bus services town maps are being developed to show those roads that will not normally have any form of traffic calming measures installed and those which, if traffic calmed, will not incorporate vertical deflections. Nottinghamshire is adopting a similar approach following a detailed review of traffic calming policy and practice.

Some of the authorities who use vertical deflections only as a last resort on bus routes hope to extend this policy to other roads following complaints from the local residents who have to negotiate the humps most frequently. Such an extension will have to be carefully assessed in relation to individual scheme safety objectives. On the other hand there can be circumstances where very long plateau style humps may be appropriate on bus routes. For example in shopping streets as part of some environmental enhancement measures; these streets may also be suitable for bus priority measures.

Defining roads which will not be traffic calmed does not mean introducing a new or different road hierarchy system; it builds on the established hierarchy. Nor does it mean that other traffic management mea-

sures will not be introduced where appropriate on any road within the hierarchy. Early discussions with the relevant local authority public transport co-ordination staff will provide overall advice on bus services to assist in developing draft proposals.

The emergency services and bus operators have found the road hierarchy approach to traffic calming constructive. It removes their fear that every road in a town will eventually have humps installed and it produces key routes for the emergency services to use when answering call-outs. As UTC traffic light control schemes come forward the same key routes can form a basis for planning signalised 'Green Wave' systems. The aims are to allow the emergency services to travel to within acceptable distances of local housing, commercial and shopping areas, with only the final lengths of journeys affected by traffic calming incorporating road humps, and for buses to travel on roads that they use frequently without having to negotiate road humps.

At present there is insufficient data to give firm advice on the options that should be considered for all towns. The following are therefore tentative suggestions using the limited information available.

Having established the road hierarchy for a town[52] roads may be categorised for :

- no traffic calming measures within the road although other traffic management techniques can apply;
- no traffic calming measures closer than 100m intervals along the road with a presumption against road humps;
- no 'standard' road humps, no chicanes of stagger length less than 17m and 3.5m lane width along the road;
- roads suitable for any traffic calming measures, if needed to solve problems.

The preparation of traffic calming town plans in consultation with the bus operators and emergency services is not simple and requires all parties to recognise that there are conflicts of interest. One authority invited the fire and ambulance services to submit their ideas before preparing its own proposals. The ambulance service will tend to seek a finer network than the fire service. However, there should be enough common ground between them, the bus operators and the highway authority to find a workable solution *provided each acts reasonably by adopting a corporate approach which recognises the others' needs.* A good example of this is the joint statement by the emergency services in Kent's latest code of practice on traffic calming.[75]

An alternative approach to categorising roads to define the extent of traffic calming which is appropriate is the 'Hierarchy of Uses' approach, devised for the South Birmingham Environmental Traffic

Management project (SOBETMA). This was developed for a major radial urban road corridor, and involved defining for each section which uses were of most importance — eg, shopping, residential, or through traffic — and designing road space allocation, speed reductions and priority for crossings according.[84]

Strategic traffic calming town plans do not remove the need for early consultation with bus operators and the emergency services on individual schemes as outlined below. Nor should the lack of an overall plan preclude worthwhile individual schemes going ahead in any area, urban or rural. But care must be taken to ensure that an ad-hoc approach does not cause incremental problems to develop.

Where possible, cyclists should be provided with cycle lanes or cycle slips which enable them to bypass such physical restrictions as chicanes and narrowings. Picture: Laverstock, Church Street (Case Study 85).

Vulnerable Road Users

Generally, lower vehicle speeds will benefit all types of vulnerable road user, but it is important to consult local representative groups to ensure that the particular concerns of each are taken into account. Vulnerable road uses include:

- pedestrians, especially children and the elderly or infirm;
- people with the different forms of mobility difficulty, including visual handicaps;
- cyclists, motorcyclists and other non-motorised wheeled road users.

Often relatively simple low cost action may improve accessibility for those who have to use wheelchairs or whose sight is impaired, and this will usually also mean an improvement in conditions for all pedestrians. Practical advice can be found in the IHT Guidelines for Reducing Mobility Handicaps.[57]

Most authorities have, or are developing, cycle networks for built up areas. These fit in well with the road hierarchy concept and enable a strategic approach to be taken when considering the effects on cyclists of traffic calming plans. Although in general cyclists benefit from lower traffic speeds, poorly designed schemes can inconvenience or even endanger cyclists. Cyclists will not be attracted to routes which prevent them building up and maintaining momentum.

Where possible, cyclists should be provided with cycle lanes or cycle slips which enable them to bypass such physical restrictions. Without these, buildouts and chicanes in particular may force the cyclist into conflict with motor vehicles, creating a risk of collision. The minimum recommended width for a

cycle pass is 700 mm. The Cyclists Touring Club has published a technical note on traffic calming [44] which provides valuable advice for designers. Local representatives of the CTC or other cycling clubs should be among the list of consultees for all traffic calming schemes. Fourteen of the case studies included in this book highlight specific measures to assist cyclists; these are listed in Table 1, on page 25.

Designers are generally aware of the problems with traffic calming presents to motorcyclists — and of the way that some irresponsible motorcyclists react to the 'challenge' of speed restricting measures. If there are local organisations which represent this group of users then their advice should be sought at the strategic level as well as on individual schemes.

Consultation: The Statutory Minimum

Statutory requirements to advertise speed limits are set out in Regulation 6 of the Local Authorities' Traffic Orders (Procedure) (England and Wales) Regulations 1989, and requirements to advertise plans for road humps in Section 90C of the Highways Act 1980. In addition, Regulation 90C(1) of the Highways Act 1990 and Regulation 3 of the Highways (Road Humps) Regulations 1990 require formal consultation on proposals for road humps with the police and one or more organisations representing users. Regulation 4 of the Highways (Traffic Calming) Regulations requires the police to be consulted, and such other persons as the highway authority thinks fit, on traffic calming works as defined by the regulations.

The Department of Transport recommends more extensive consultation. Its Circular Roads 3/90[7] urges consultation on road humps proposals with the local fire brigade, ambulance service and bus operators. The Department's Traffic Advisory leaflet on 20 mph zones[21] recommends wide consultation on proposals at an early stage with a range of organisations including bus operators, the police, fire and emergency services. The advice is that "This should be before the zone area has been determined, and at a stage when options for the type of speed restraint facilities that might be used are still open." The Department's procedures for the submission of applications for 20 mph zones require the inclusion of responses from the emergency services and, if relevant, bus operators regarding the proposed schemes.

The above can be regarded as the minimum necessary but the designer of any scheme is urged to follow the recommended approach described next.

Consultation: The Recommended Approach

Authorities should adopt an open and fully consultative approach to the formulation and implementation of all traffic calming proposals. *Consultation means being prepared to change not only details but basic concepts if cogent arguments are put forward.* A 'partnership dialogue' should be created between all interested parties. Interested parties would include residents associations, groups representing people with visual and physical difficulties, cycling groups, Chambers of Commerce, taxi firms, commercial groups and road freight haulage associations as well as the emergency services and bus operators.

The recommended process for promoting a traffic calming scheme is shown in Figure 1 on page 14, which shows the stages at which informal and formal consultation should take place throughout the promotion process.

In London, the Bus Priority and Traffic Unit on London Transport Buses has suggested that initial discussions should be held with them as soon as the need for traffic calming is identified on an existing or potential bus route[72]. This is sound advice and early contact should be made with all the relevant bus operators, ideally using the road hierarchy approach described above.

As well as consulting the local authority section dealing with public transport co-ordination the designer should also involve colleagues in other departments of the authority, particularly planning and education. Close liaison will enable knowledge about other aspects of the physical planning of the community to be made available. Local schools could participate in a project to contribute ideas. This will help to take account of children's views and give them a practical application for a curriculum subject.

As stated earlier each authority will have its own preferred methods of local public consultation which will vary to suit the circumstances of a particular proposal. Several authorities use local liaison groups with representatives from interested parties. Most hold public meetings and exhibitions, recognising that the former are not always representative of local opinion. Questionnaires are frequently used, although asking the right questions is not straightforward. All methods require considerable resource input, often at senior level, and whilst meaningful public consultations are necessary the resource implications should not be underestimated and should be kept in scale with the project and its objectives.

Where special authorisations are needed sufficient time for the process must be allowed. The time can vary, depending on the nature of the authorisation, but at least 12 weeks should be assumed.

Clarity of Purpose

There is intense public pressure on many highway authorities to introduce traffic calming measures. The consultation process must try to ensure everyone appreciates the advantages and disadvantages of any measures so that they are not assumed to be a cure for all problems. The appropriateness of the measures for each type of road needs careful consideration.

It is important that emergency services are not delayed unduly and that bus operators continue to provide services. The traffic engineer has a difficult task in weighing up all the pros and cons and giving advice to Members accordingly. The key to a successful decision, however, is a clarity of purpose, backed up by an understanding of local issues and views discovered through well planned consultations and by being prepared to change the scheme when necessary.

CHAPTER 3

Comparative Overview of the Case Studies

The 85 case studies which form the core of this book are intended to provide a snapshot of the experience in traffic calming which has accumulated in the UK. The cases have been selected to give a reasonable cross section of schemes and types of measures applied. The intention is to allow practitioners to identify schemes with similarities to the locations which they themselves are dealing with, in order to draw lessons from the types of measures used and the effects these have achieved in practice. The preponderance in this book of schemes for residential streets reflects the fact that across the country traffic calming has been most often applied in such areas.

This chapter comments on some of the patterns which emerged from examining all the 152 cases originally submitted for possible inclusion in this book. A more detailed overview of the 85 schemes selected for inclusion is provided by the tables at the end of this chapter. Table 1 is an index to which case studies have made use of the different types of traffic calming measures. Table 2 aims to provide an 'at a glance' summary of monitoring data, relative costs and the apparent effectiveness of the case study schemes.

The body of this book, which follows, is the illustrated case studies themselves. Each sets out a necessarily brief description covering background information, design details, and before and after data on traffic, accidents and speed, following a fairly standardised layout to assist readers in making their own comparisons. The details given should allow readers to decide whether schemes may provide lessons relevant to their own areas, and a contact name and telephone number is given for follow up purposes. In most cases the local contact's own brief views about the case are also included.

Work on this project started with an invitation to local authorities, circulated through their appropriate technical associations, to send in examples of traffic calming schemes that had been constructed in their areas. Both successful and unsuccessful cases were sought so that, in the latter case, similar schemes could be avoided or modified. We are particularly grateful to those authorities who have been willing to allow us to publish schemes which were, initially at least, less than successful and to explain how these have subsequently been modified. In practice, however, few truly unsuccessful schemes were submitted.

Patterns Emerging from the Cases Submitted

The numerical analysis which follows draws on all the 152 case studies submitted, and highlights some points of interest.

Carriageway narrowings are popular, and one of the lowest cost case studies used road markings to tackle 3 km of dual carriageway: Plymouth, Crownhill Road (Case Study 56).

- **Location Types:**

   Just over half the schemes (56%) were located in urban residential areas with 29% occurring on main roads in rural locations and the remaining 15% in town centre areas.

- **Scheme Objectives:**

   74% of the schemes were aimed at reducing speed, with 64% aimed at accident reduction, and 36% at reducing the volume of through traffic. Nearly every scheme was intended to address two of these problems and many were aimed at all three. Environmental improvements were sought in 17% of the schemes, fairly evenly spread over the three types of location. This low proportion is probably related to the relatively high costs associated with environmental improvements.

- **Costs:**

   60% of the schemes fell within the cost range £10,000 to £100,000. About 20% cost less than £10,000 and the same percentage cost over £100,000. Significantly 9% of the schemes cost more than £250,000, so traffic calming should not be seen as a cheap measure. The costs of successful measures ranged from about £3,000 for Tuly Street, Barnstaple, Devon (Case Study 6), to over £2000 per metre of road length treated as in Carfax, Horsham, West Sussex (Case Study 1). The latter was linked to major town centre redevelopment and in common with the other relatively expensive schemes incorporated significant environmental enhancements using high quality materials.

   Several of the schemes were reported have been funded by partnerships between different local authorities and in some cases with contributions by developers, particularly for town centre schemes (see Case Studies 1, 3 and 6). A contribution from the community is reported for Fen Ditton, Cambridgeshire (Case Study 20).

   Cost comparisons between different case studies need to be treated with caution. A full comparison would consider traffic volume as well as road length treated, and would have regard to the degree of effectiveness in environmental improvement as well as road safety terms. However it is readily apparent that though successful traffic calming is seldom cheap, there can be occasions when quite a modest measure can be effective. Designing for a combination of traffic calming and environmental enhancement is probably ideal if funding can be found. However sometimes an effective traffic calming scheme can be achieved without linking it to environmental measures, and judgements to balance need, priority and resources have to be made.

Techniques Employed

76% of the schemes submitted used more than one type of speed controlling device. Some schemes involved a whole range of different types of measure. Road humps in one form or another were included in 66% of the schemes submitted and were by far the most commonly used technique. However, twelve of the Rural schemes, nine of the Residential and two of the Town Centre case studies included here were 'no hump' schemes; these are identified in Table 1.

The next most popular technique was some form of carriageway narrowing, with 41% of schemes using this method. Chicanes featured in 26% of schemes and, below this, 17% of the schemes included refuges, mini roundabouts, road markings and signs (as a main feature) and environmental improvements of one kind or another. Other relatively commonly used measures were gateways (10%), table junctions (9%) coloured surfacing (7%) and rumble strips (6%). Protected parking areas featured in 13% of the cases.

Only 8% of schemes submitted were 20 mph zones but this is probably because few of the 100 now in existence have yet been in place long enough for performance to be monitored.

Measures employed only rarely were road closures, parking restric-

Gateways are becoming common features at the approach to villages, but measures within the village are usually needed as well. Picture: Devon, Newton Tracey (Case Study 21).

tions, changes to junction priorities, speed cushions, visual effects and speed cameras. Of course speed cushions only began to be more widely used in the UK in late 1993 and although quite a large number of schemes using them now exist they have not been in place long enough for assessment. Similarly speed cameras have only started to appear relatively recently, and one such scheme is included, Nuneham Courtenay in Oxfordshire (Case Study 17).

However, it appears that some authorities may not be considering the potential for use of more traditional traffic management techniques, such as road closures, priority at junctions and parking controls, in combination with traffic calming.

There is potential for use of traditional traffic management measures, such as road closures, in conjunction with traffic calming. Picture: Newham, London (Case Study 78).

Effectiveness

Effectiveness should be assessed against scheme objectives which (as noted in the preceding chapters) can vary greatly, from sites with a known poor accident record requiring attention, to places where residents perceive danger or simply object to the speed and/or the volume of traffic passing. In nearly all the cases submitted it appeared that there had been a measurable problem of accidents, speed or traffic flow which prompted action.

The effectiveness of the measures employed seems to vary considerably and it is not certain whether this is the result of differences in design or of location. From the schemes sent in was it not possible to identify any one specific measure as more or less effective. Effectiveness, it seems, will depend on the way measures are fitted to local conditions and on the combination of measures used. It is not always related to cost, as much of the cost on the more expensive projects is due to the use of high quality materials for carriageway and footway reconstruction and environmental treatment rather than to the functional backbone of the scheme.

Only four of the schemes submitted were described as failing to achieve what was intended, though it may be significant that no results were submitted for a further 30 schemes. These four have all been included as case studies and are West Bretton, West Yorkshire (Case Study 33), Matfield, Kent (Case Study 16) which is one of the VISP projects[83]; Selsey, West Sussex (Case Studies 37-38); and Christchurch Road, Bexley in London (Case Study 60). Twelve schemes were considered to be only a partial success whilst 104 (almost 70%) were described as having achieved their aims.

20 mph Zones

As noted above, there are now about a hundred 20 mph zones in operation and nine of these are included as case studies. Two are in Town Centres: Carfax, Horsham, West Sussex (Case Study 1) and High Street, Rushden, Northamptonshire (Case Study 25), both of which feature extensive environmental enhancements. The former was carried out as part of a major redevelopment scheme managed by the local authorities.

Road humps or plateaus are common features in all the seven Residential 20 mph zones published here, as well as in the two Town Centre cases. The changes in vertical profile are usually in combination with fairly rigorous carriageway width restrictions in the form of chicanes, footway build outs or narrowings, and occasional road closures.

The zones all seem to have reduced personal injury accidents considerably, but it is not surprising to see that area-wide application of severe vertical and horizontal displacement measures has been needed to reduce vehicle speeds sufficiently to achieve this scale of casualty reduction. Some authorities are currently trying to establish 20 mph zones which do not incorporate road humps.

Speed Cushions

The ongoing TRL research into speed cushions has already been mentioned in Chapter 2. Three schemes using cushions are included here: Victoria Road, Huddersfield, West Yorkshire (Case Study 70); Cedar Road and Queen Elizabeth Road (Case Studies 44 and 45), both part of the Camp Hill 20 mph zone in Nuneaton, Warwickshire. In all these three cases, use of the cushions is integrated with one-way working pinch points, and appears to be effective.

Monitoring

In nearly all the case studies published here fairly comprehensive monitoring of traffic flows, speeds and accidents has taken place. But surprisingly it does seem that some of the schemes submitted had not been monitored. The vital importance of monitoring all schemes is emphasised in Chapter 1, and must be stressed again here.

Consultation

The extent and methods of consultation used were varied. In many locations extensive prior public consultation had taken place in the form of public meetings and exhibitions. Liaison groups had been used in some cases. In a few cases consultation had been simply the minimum statutory consultation and on occasion it appeared that bus operators had not been included in the consultations.

There is no doubt that prior consultation can improve the public acceptability of schemes, but it can also have the side effects of precipitating and focusing opposition that may not be justified or which comes from unrepresentative sources. The overall message, as stressed in both the preceding chapters, is that consultation needs to be handled extremely carefully and it must be undertaken properly.

Key Points:

The following points have been drawn from all the 152 case studies submitted not just those included in this book.

- Some designers appear to focus solely on traffic calming measures rather than using traditional traffic management and traffic calming measures in combination.

- Road humps are an effective means of speed reduction, but are often opposed by bus operators and the emergency services. In some situations it should be possible to achieve a sufficiently effective scheme without the need for vertical deflections.

- Ramp gradients vary considerably, but the range between 1:10 and 1:15 combines the greatest effectiveness with the least harshness.

- Whilst road humps are effective they invariably attract criticism for the inconvenience, discomfort and alleged vehicle damage that they cause.

- 'Platforms' or 'tables' seem more acceptable to drivers than 'standard' humps and can be adequately effective in speed reduction.

- Rumble strips and jiggle bars have negligible effect on speed but can be effective driver alerting devices. However the noise made often causes residents to complain and leads to the eventual removal of the strips.

- Chicanes can work well at slowing traffic; their effectiveness is related to the severity of the geometry. Care must be taken when access is required for larger vehicles or on bus routes.

- 'Throttles' or narrowings can also be very effective particularly when two-way vehicle flows are relatively high so that greater caution is required. However, they may reward aggressive driver behaviour. They do need adequate signing and marking in advance to avoid becoming a safety hazard in themselves.

- 'Gateway' treatments tend to be ineffective unless they incorporate some form of physical narrowing including vertical design elements or a chicane.

- The effect of gateways can be short-lived and requires repeater features within the village or urban road to maintain speeds at a lower level.

- The introduction of a mini-roundabout or changed junction priorities can have a 'calming' effect, particularly when linked to other features.

- The dome height at mini-roundabouts needs careful design where there are right turning buses in a restricted space.

- Combinations of measures often work well together where individually they may have little effect.

- Some authorities have been particularly innovative in designing calming features and there is considerable scope for experimentation in this field. Great care must be taken as public safety is involved and it is important to seek legal advice and to check whether DOT special authorisation is needed before implementing a new feature.

- Speed cameras have been used for their deterrent effect on slowing traffic. They seem to be effective, but do require enforcement resources and do not have a noticeable effect on deterring rat-running traffic.

It is very clear that modifications are being made to traffic calming measures already in use and entirely new measures are being developed continuously. As experience grows, provided systematic monitoring takes place, it will be possible for authorities to decide which measures are appropriate for different circumstances. Ultimately it will be for designers to make their decisions having regard to all the local circumstances.

TABLE 1. MEASURES USED IN THE CASE STUDIES

TYPES OF MEASURES	Town Centre	Rural	Residential
20mph Zones	1, 2		39, 40, 41, 42, 43, 44, 45, 47
Chicanes	2,4,10	18, 20, 26, 29, 32, 35	39, 40, 46, 51, 55, 57, 62, 64, 66, 69, 76, 80, 82, 83
Cycle Measures: Slips, Lanes, 'Bypasses'	4	20, 27, 34	45, 46, 52, 53, 56, 61, 63, 78, 83, 85
Environmental Improvement	1, 2, 3, 5, 9,13	25, 26, 29, 35	40, 53, 54, 55, 65, 75, 77, 78
Gateways	1,13	15, 16, 17, 18, 21, 23, 26, 29, 30, 32, 37, 38	39, 51
Junction Priority Changes	1, 7	22	42, 57, 62, 75, 77, 83
Junction Treatments: Tables, Entry/Exit	11,12	17, 22, 25	39, 41, 43, 48, 49, 53, 64, 65, 68, 78
Mini-Roundabouts	3,4,11,12	22, 24, 25, 26, 28, 31, 36	43, 54, 55, 58, 63, 65, 66,
Narrowings: Build outs, Pinch points, Refuges	1, 2, 3, 4, 5, 6, 7, 8, 10, 13	14, 15, 18, 19, 20, 21, 22, 23, 24, 26, 27, 28, 29, 30, 32, 34, 35	39, 40, 42, 43, 44, 45, 49, 52, 53, 55, 56, 57, 58, 62, 63, 64, 66, 67, 68, 69, 70, 75, 76, 77, 78, 80, 85
Parking: Protected, Restrictions	2, 5, 8	24, 25, 29	39, 40, 43, 52, 53, 55, 57, 64, 66, 69, 83
Road Closures	1	26	39, 42, 61, 63, 73, 78
Road Humps: Flat Top, Round Top, One-Way, Two-Way, Plateaus Speed Tables	1, 2, 3, 5, 6, 7, 8, 9, 10, 12	19, 22, 24, 25, 28, 31, 35, 36	39, 40, 41, 42, 43, 46, 47, 48, 49, 50, 53, 54, 56, 57, 58, 59, 62, 63, 65, 67, 68, 71, 73, 74, 75, 76, 77, 78, 80, 81, 84, 85
Rumble Strips: 'Thumps', Mini-humps		18, 23, 32, 33, 37	56, 60, 70, 71, 72, 76, 80, 84
Speed Cameras		17	
Speed Cushions			44, 45, 56, 70
Surface Treatments: Colour, Texture	1,13	15, 21, 23, 24, 29, 38	52, 78
Visual Effects: Signs, Markings, Planting	13	15, 16, 18, 19, 21, 27, 30, 32, 34, 38	43, 52, 56, 65, 69, 80, 83
'No Humps': Successful schemes	4, 13	14, 15, 17, 18, 21, 26, 27, 29, 30, 32, 34, 38	51, 52, 55, 56, 61, 66, 69, 70, 82

QUICK REFERENCE GUIDE TO THE CASE STUDIES

Table 2.1 Town Centre Case Studies

Case Study	Location	Traffic volume Before	After	Speed mph Before	After	Accidents pia pa Before	After	Cost	Comments (See Note 1 below)
1	West Sussex Horsham Carfax	-	-	35	15	3	0	£2.3m	20 mph zone. Linked with redevelopment and town relief road. Very expensive environmental works. Considerable accident and speed reduction.
2	Northamptonshire Rushden High Street	2540	2080	24	17	3.7	2	£230k	20 mph zone (temp). Flat top humps, pinch points, chicanes. Expensive environmentally. Speed drop significant, considerable (potential) accident drop.
3	Hertfordshire Borehamwood Shenley Road	18800	16500	26	20	15	8	£1.2m	Major enhancement. Narrowing, road humps. Developer funding. Relatively expensive, significant speed reduction, considerable accident reduction.
4	Cambridgeshire Eaton Socon	10700	8900	38	32	19	4	£130k	Mini roundabouts, traffic islands, chicanes, cycle lanes. Reasonable cost. Significant speed reduction, considerable accident reduction.
5	Cornwall Saltash Fore Street	9000	7900	32	17	6	1	£230k	Road humps, narrowing, protected parking. Environmental treatment. Relatively expensive. Considerable speed and potential accident drop.
6	Devon Barnstaple Tuly Street	1800	4000	23	11	0.25	0.25	£3k	Combined narrowing with flat top hump. Developer financed. Low cost with considerable speed drop. Extra traffic due to short term car park.
7	Essex Southend on Sea Marine Parade	-	-	28	26	28	9	£181k	Junction alterations, narrowing, flat top humps, Pelicans. Relatively expensive. Considerable accident and pedestrian injury drop. More extensive consultation may have helped.
8	Kent Gillingham Twydall Green	5000	5000	-	-	1.3	0.7	£18k	Early road hump scheme on dual carriageway. Relatively expensive but considerable accident reduction.
9	Kent Rochester Vines Lane	8000	4000	35	25	2	0	£18.5k	High quality single way ramps used to slow traffic and deter rat running. Reasonable cost with considerable accident and speed reduction.
10	Isle of Wight Ryde North Walk	850	815	29	18	0	0	£14.5k	Pinch points plus humped chicane. Reasonable cost scheme with considerable reduction in speeds in a recreational area well used by pedestrians.
11	Warwickshire Leamington Spa The Parade	15600	17500	-	20	3	5	£35k	Mini roundabouts on speed tables. Expensive. Special paving had to be replaced. Not successful. Traffic patterns changed due to other factors.
12	Warwickshire Rugby Ashlawn Road	7100	5400	50	25	2	0	£69k	Raised junction plus several flat top humps. Relatively expensive but considerable speed and accident reduction.
13	Hampshire Petersfield Dragon Street	NYA		NYA		NYA		£480k	DOT demonstration project. Detrunked 'high street' location. Narrowing, extensive environmental works. Intensive consultation. Expensive.

Note 1. Comparative subjective comments are given regarding cost and effectiveness. These are based on the following indicative criteria which may differ from individual authorities' own criteria and opinions:

COST	'Low'	'Moderate'	'Reasonable'	'Expensive'
Single Measure	<£3,000	£3,001 – £10,000	£10,001 – £20,000	>£20,000
Several Measures	<£,5000/km	£5,001 – £50,000/km	£50,001 – £100,000/km	>£100,000/km

EFFECTIVENESS	'Negligible'	'Marginal'	'Significant'	'Considerable'
Accidents reduced by	<5%	5% – 10%	10% – 30%	>30%
Speeds reduced by	<2 mph	2 – 5 mph	5 – 10 mph	>10 mph

Table 2.2 Rural Case Studies

Case Study	Location	Traffic volume Before	After	Speed mph Before	After	Accidents pia pa Before	After	Cost	Comments (See Note 1, p. 27)
14	Somerset Stratton-on-the-Fosse	4000	5000	47	35	1.8	1	£50k	VISP scheme. Temporary materials, trial period. Narrowings priority features. Cost moderate with considerable speed drop.
15	Devon Halberton	2000	2000	44	36	0.7	1.5	£9k	VISP scheme. Gateway features, traffic islands, markings and speed roundels. Moderate cost with significant speed drop.
16	Kent Matfield	6900	6900	40	38	2.3	2	£ 2.2k	VISP low cost scheme involving specially authorised gateway signing. Not particularly effective.
17	Oxfordshire Nuneham Courtenay	17500	17500	47	41	1.3	1	£12k	Speed camera. Reasonable cost (excludes fines income). Significant speed drop. Little effect from gateways added later.
18	Hampshire New Forest	See	case	study				£400k	40mph zone (300 km road). Comprehensive approach to Heritage Area. Accidents and speeds reduced by range of features.
19	Cambridgeshire Gamlingay	1700	1500	37	27	2.7	0.3	£32k	Give way narrowings, series of round and flat top humps, bus route. Modest cost with considerable accident and speed reduction.
20	Cambridgeshire Fen Ditton	10600	11300	47	41	5.7	2.7	£48k	Traffic islands and chicanes with 'by-passes' for cyclists. Moderate cost with considerable potential accident reduction and significant speed drop.
21	Devon Newton Tracey	2900	2900	38	35	0.3	0	£20k	Gateway effect with central island located on entry to village. Reasonable cost but marginal speed drop.
22	Hertfordshire Buntingford	7300	6850	35	24	2.8	1	£400k	3½ years consultation. Narrowing, speed tables, throttle humps, mini. High quality enhancement. Expensive. Considerable speed and accident drop.
23	Hertfordshire Old Knebworth	2000	1800	42	39	0.7	0	£18k	Experimental on 60mph road. Narrowings, castellated surfacing, gateways. Moderate cost. Negligible speed drop. More measures proposed.
24	Hertfordshire Stanstead Abbots (1)	7100	7100	31	16	2	-	£82k	Mini roundabout, road humps, narrowing. Relatively expensive. Speed and accident drop considerable. Experimental scheme, replaced by Case Study 25.
25	Hertfordshire Stanstead Abotts (2)	7100	7000	31	22	2	-	£476k	Extensive environmental works. Replaces Case Study 24. Humps replaced by raised table areas. Expensive but considerable speed and accident drop.
26	Kent Sarre	10000	10000	47	34	3.7	0	£140k	Two primary routes in village. Environmental works. Gateways, refuges, chicane, mini, closure. Expensive but considerable speed and (potential) accident drop.
27	Kent Gravesend	9400	9400	60	53	5	2.4	£75k	Reduced number of lanes, refuges, cycle lanes. 2.5 km. Cost moderate. Significant speed drop and considerable accident drop.
28	Kent Hoo St Werburgh	8500	9800	45	27	5	0	£300k	Extensive scheme. Flat top humps, minis, refuges. Landscaping. Joint funding. Expensive but speed and (potential) accident drop considerable.
29	Kent Brasted	13500	13500	40	35	3	2.4	£425k	Gateways, chicanes, textured surfaces and street lighting. Expensive environmental works with marginal speed drop.

Table 2.2 Rural Case Studies (cont)

Case Study	Location	Traffic volume Before	After	Speed mph Before	After	Accidents pia pa Before	After	Cost	Comments (See Note 1, p. 27)
30	Leicestershire A427, Five Villages	10200	-	51	44	10.7	8	£50k	Villages on a principal road. Narrowings, Gateways, lighting improvements, signs, markings. Low cost. Noticeable speed drop.
31	Oxfordshire Kennington	8400	6300	37	28	4	1.7	£55k	Rat run. Road humps, minis. Moderate cost with significant accident and speed drop. Lower humps may have been OK.
32	Strathclyde Croy	550	550	49	41	3.7	1	£15k	Gateways, refuges, rumble strips, signs, markings. Moderate cost. Significant speed drop in one direction and considerable accident drop potential.
33	West Yorkshire West Bretton	-	-	48	39	-	-	£2.5k	Low cost jiggle bars. Unsuccessful. High and low speeds in conflict and noise complaints from residents. Jiggle bars removed.
34	West Sussex Birdham	11350	11350	56	44	4.3	4.6	£50k	Measures include narrowing, the provision of a cycle track and hatched central reserve. Moderate cost with considerable speed drop.
35	West Sussex Bramber	5000	3600	-	-	0.3	0	£85k	Environmentally sensitive. Road humps, narrowing, chicane. Expensive but main objective to reduce through traffic achieved.
36	West Sussex Fontwell	1300	1080	40	25	0.3	0	£25k	Road humps, minis, 30 mph limit, to encourage through traffic onto by-pass. Moderate cost with considerable speed drop.
37	West Sussex Selsey (1)	10100	10100	42	39	1.3	0.8	£9.5k	Combination of gateways and mini road humps. Unsuccessful. Noise complaints from residents. Replaced by Case Study 38.
38	West Sussex Selsey (2)	10100	10100	40	32	1.3	0	£3k	Edge and centre markings and road studs to narrow carriageway. Low cost with significant speed drop. Replaced Case Study 37.

Table 2.3 Residential Case Studies

Case Study	Location	Traffic volume Before	After	Speed mph Before	After	Accidents pia pa Before	After	Cost	Comments (See Note 1, p. 27)
39	North Yorkshire York The Groves	4000	4000	27	15	2.3	1	£130k	20 mph zone. Gateways, humps, build outs, one way plugs, chicanes, raised junctions, residents parking. Moderate cost. Considerable speed and accident reduction.
40	West Yorkshire Bradford Scotchman Road	680	600	32	19	1.3	1	£105k	20 mph zone. Chicanes, narrowings, platforms in two streets. Relatively expensive due to materials and street lighting but considerable speed drop.
41	Suffolk Ipswich Britannia Area	2000	2000	30	16	4.2	1	£85k	20 mph zone. 40 humps of different types. Moderate cost with considerable drop in accidents and speeds.
42	Manchester Crumpsall Green	2500	1100	-	12	8.8	NYA	£175k	20 mph zone. Humps, signals, narrowing, one-way, closure. Reasonable cost. Full evaluation not yet available.

Table 2.3 Residential Case Studies (cont)

Case Study	Location	Traffic volume Before	After	Speed mph Before	After	Accidents pia pa Before	After	Cost	Comments (See Note 1, p. 27)
43	Dorset Poole Upper Parkstone	5333	-	27	16	25	-	£320k	20 mph zone. Plateaus, protected parking, mini - roundabouts. Relatively expensive but considerable speed drop and accident reduction potential.
44	Warwickshire Nuneaton Cedar Road	3000	3000	30	20	2.7	0	£8k	20 mph zone. 4 pinch points with speed cushion. Low cost and considerable speed reductions. No accidents to date. Compare Case Study 45.
45	Warwickshire Nuneaton Qu. Elizabeth Road	2500	2500	42	27	4	-	£10k	20 mph zone peripheral road. 5 offset pinch points with speed cushion. Low cost, considerable speed drop. No accidents to date. Compare Case Study 44.
46	Warwickshire Nuneaton Whittleford Road	9000	9000	41	30	11	1	£4k	20 mph zone peripheral road. Severe chicanes now removed due to complaints about congestion. Low cost with considerable speed reduction.
47	Staffordshire Stoke on Trent Fenton Area	350	220	40	15	1.5	0	£79k	20 mph zone. Flat top humps in block paving . Some humps modified for hearses. Moderate cost. Considerable accident and speed reduction.
48	Avon Bristol Throgmorton Road	7000	3750	36	23	2.8	0	£40k	Flat top humps and raised junction on a bus route. Ramps to be extended. Reasonable cost with considerable speed and potential accident drop.
49	West Yorkshire Bradford Upper Rushton Rd	3400	2000	34	25	4.3	0	£70k	A series of single way narrowed ramps, entry ramps and junction table. Reasonable cost with considerable speed and accident drop.
50	London West Wickam Hawes Lane Area	700	530	39	21	2.7	1.5	£31k	Round top humps. Moderate cost includes additional measures for displaced traffic. Speed drop and potential accident reduction considerable.
51	Clackmannanshire Tillicoultry Stalker Avenue	-	-	30+	21	0.8	0	£10k	Gateways and chicanes with over-run areas for buses. Moderate cost with considerable drop in accidents and speeds.
52	Devon Barnstaple Park School	7000	7000	40	25	0.2	0	£20k	To calm traffic at pedestrian crossing point using refuges, hatching and coloured surfacing. Reasonable cost with considerable speed drop.
53	Devon Exeter Burnthouse Lane	6200	5500	34	24	6.6	1.4	£220k	Series of flat top humps, high quality materials, street lighting, landscaping. Relatively expensive but considerable accident and speed reduction.
54	Devon Exmouth Withycombe Village Road	7900	7000	33	25	4.7	0.3	£91k	Flat top humps, footway extensions, protected parking, mini roundabouts. Relatively expensive. Considerable accident reduction and significant speed drop.
55	Devon Plymouth Beaumont Road	3900	3300	37	32	5.3	3	£99k	Narrowing, chicanes by widening footways and planting features. Bus route. Reasonable cost. Considerable accident and noticeable speed drop.
56	Devon Plymouth Crownhill Road	14000	14000	45	41	27	14	£12.6k	A lane and hatch marking scheme to narrow a dual carriageway and provide cycle lanes. Low cost and considerable accident drop.
57	Dumfries & Galloway Dumfries Calside Road	4000	4000	35	25	0.3	0	£36k	Flat top humps, narrowing, refuges, protected parking. Bus route. Moderate cost. Considerable speed drop. Potential hazards at schools.

Table 2.3 Residential Case Studies (cont)

Case Study	Location	Traffic volume Before	Traffic volume After	Speed mph Before	Speed mph After	Accidents pia pa Before	Accidents pia pa After	Cost	Comments (See Note 1, p. 27)
58	Essex Harlow Parringdon Road	-	-	33	27	30	15	£77k	Flat top narrowed humps, priority working, mini. Some driver conflict. Moderate cost. Significant speed drop, considerable potential accident drop.
59	Essex Hadleigh Scrub Lane	8700	3580	40	27	3	0	£46k	An extensive scheme of round top road humps. Reasonable cost with considerable speed reduction and potential accident reduction.
60	London Bexley Christchurch Road	-	-	36	31	-	-	£9.3k	Low cost experimental installation of thermoplastic rumble strips. Rejected by a majority of residents and subsequently removed at £24k cost.
61	Gloucestershire Gloucester Conduit Street	-	-	-	-	-	-	£10.5k	A series of road closures by footway extensions to prevent speeding and rat running. Wide local support. Moderate cost.
62	Gwent Newport Llanthewy Road	2230	1830	32	20	0.7	0	£25k	An entry chicane plus road humps and narrowing. One-way street. Moderate cost with considerable speed drop.
63	Hertfordshire Hemel Hempstead Peascroft Road	720	270	44	18	5.4	0.5	£52k	Round and flat top humps, mini roundabouts, priority narrowing. Reasonable cost and considerable accident and speed reduction.
64	London Shepherds Bush Wulfstan Street	820	450	35	20	4.7	0.7	£70k	Speed table, chicanes, width restrictions, refuges and junction treatments. Reasonable cost with considerable speed and accident drop.
65	South Humberside Grimsby Grant Thorold Park	-	-	-	-	16.7	7.5	£122k	Entry treatment, speed tables and mini round abouts all in quality surfacing. Moderate cost with considerable accident reduction.
66	South Humberside Grimsby Nunsthorpe Estate	-	-	-	-	5.3	2	£201k	Environmental enhancement. Chicanes, refuges, protected parking, mini roundabout. Relatively expensive but considerable accident reduction.
67	Isle of Wight Newport Pan Estate	4830	3180	35	17	3.4	1.7	£40k	Numerous flat top road humps, plateaus incorporating bus lay bys. Reasonable cost with considerable accident and speed drop.
68	Kent Sittingbourne Stanhope Avenue	5050	3500	39	29	4	1	£43k	Block paved plateaus with narrowing features. Moderate cost with considerable accident and speed reduction.
69	Kent Tonbridge Brook Street	7000	6500	44	37	2.7	1	£42k	Unusual scheme of 'build-outs' and refuges to produce chicanes. Reasonable cost with noticeable speed reduction and potential accident reduction.
70	West Yorkshire Huddersfield Victoria Road	4000	4000	39	24	7	1	£80k	Formidable pinch points with speed cushions. Rumble strips. Relatively expensive but considerable accident and speed reduction.
71	West Yorkshire Golcar Sycamore Avenue	5000	5000	35	22	3	0.4	£80k	Plateaus, thermoplastic 'thumps'. Latter effective. Bus problems with plateau ramps. Relatively expensive. Considerable accident and speed drop.
72	West Yorkshire Paddock Church Street	7000	5500	35	24	8	0.5	£45k	Unusual rounded rumble strips built despite traders and bus operators opposition. Moderate cost. Considerable speed and potential accident drop.

Case Study	Location	Traffic volume		Speed mph		Accidents pia pa		Cost	Comments (See Note 1, p. 27)
		Before	After	Before	After	Before	After		
73	West Yorkshire Ravensthorpe North Road	8500	6500	35	22	3.3	1	£80k	A series of raised plateaus plus new street lighting. Relatively expensive but considerable accident and speed drop.
74	Staffordshire Featherstone The Avenue	3600	2800	39	29	2	1.2	£8.2k	Round top humps. Low cost scheme. Considerable speed reduction. Some rerouting has occurred.
75	Lancashire Lancaster Primrose Area	2600	-	34	15	2.3	-	£39k	Flat top humps with narrowings and bollards in high quality materials. Bus route. One-way streets. Reasonable cost and considerable speed drop.
76	Lancashire Burnley Brougham Street	5600	4750	35	30	6.3	2.6	£47k	Slightly raised pinch points, chicane, rumble area. Relatively expensive with considerable potential accident reduction and marginal speed drop.
77	Manchester Woodsmoor Woodsmoor Lane	2900	1650	38	22	7.7	1	£70k	Flat top humps, narrowings and priority. Bus route. Relatively expensive. Considerable speed drop and accident reduction potential.
78	London Newham Various roads	1700	90	28	0	2	0	£10k	An unusual open road closure. Raised deterrent surfacing to reinforce a limited access restriction. Moderate cost with a high degree of compliance.
79	Oxfordshire Banbury The Fairway	5700	4200	34	26	1.3	2	£17.5k	A road hump scheme with 1:12 ramps on a bus route. Modest cost. Noticeable speed drop. Accident frequency up but no child casualties.
80	Strathclyde Drumchapel Peel Glen Road	2800	1900	30	18	2.7	0.4	£15k	Combination of chicanes, narrowings, 'thumps' and markings. Moderate cost with considerable speed drop and potential accident reduction.
81	Surrey Guildford Cumberland Ave	360	240	34	19	1.3	0	£35k	A series of round top road humps. Moderate cost with considerable accident and speed drop.
82	West Yorkshire Wakefield Qu Elizabeth Road	3000	2700	38	30	1.4	0.6	£25k	Rubber kerbs and bollards for chicane system. Emergency services and bus route. Moderate cost with significant speed drop and considerable potential accident reduction.
83	Surrey Woking Albert Drive	6300	5800	**	**	12	4	£84k	Separate cycle track, chicanes (bus trials), refuges, protected parking. Relatively expensive. Speed and accident drop considerable. **See Case Study for details.
84	West Yorkshire South Elmsall Ash Grove	3000	2600	33	25	1	0.7	£13.5k	Thermoplastic 'thumps' to minimise effects on buses (special authorisation). Moderate cost. Noticeable speed drop. Replacing with chicanes.
85	Wiltshire Laverstock Church Road	4000	4400	39	25	1.1	0	£22k	Humps, cycle slips. May modify – noise complaints, congestion. Fixing problems with rubber humps. Reasonable cost. Considerable speed drop.

Town Centre

Case Studies 1 – 13

1 20 MPH ZONE, PEDESTRIANISATION
Horsham, Carfax

Location: West Sussex: Horsham, Carfax.

Implemented: April 1992.

Background: The introduction of pedestrianised streets and a 20 mph zone was a final stage in a comprehensive redevelopment of the town centre. Earlier stages included the construction of an inner relief road and shopping malls. An outer bypass also exists.

Need for Measures: Reduction in speed and volume of traffic.

Measures Installed: A 20 mph zone enforced by round and flat topped road humps. Pedestrianised areas and environmental improvements throughout.

Special Features: Extensive high quality paved areas.

Consultation: Consultation covered the period from 1985 to 1991. Initial consultation included a discussion document (January 1987), followed by a comprehensive document 'People First' (March 1989). Comments on the scheme were sought through exhibitions, brochures, newspaper articles, consultations with specialist groups, public meetings and discussions with individuals. Consultation led to the introduction of such measures as cycle racks and specially textured surfaces to assist the blind and partially sighted to avoid obstacles and locate safe crossing places.

Monitoring	Accidents (pia)	Speeds (mph)	Traffic Flow
Before	9 in 3 years	35	-
After	0 in 3 years	15	-

west sussex county council

Horsham District Council

Contact: William Knott Tel: 01243 777533 or John Parsons Tel: 01403 215444
Authority: West Sussex County Council and Horsham District Council

Technical Data:

Location Type: Town centre, part of which is designated a conservation area.

Road Type and Speed Limit: Urban unclassified: 20 mph.

Scheme Type: Round and flat topped road humps using a variety of materials within a mainly one-way 20 mph zone, incorporating road narrowings at the gateways.

Length of Scheme in Total: 1.12 km.

Dimensions:
Road humps: Height: 75 mm.
Width: Full carriageway.
Length: Plateau 3-14 m. Ramps 0.6 m.
Ramp gradient: 1:8.

Materials:
Road humps: Ramps - natural granite setts. Plateaus - York stone setts or clay paviours.
Bus and loading bays: Coloured granite setts.
Other carriageway areas: Granite setts.
Footway 'routes': Smooth sawn York stone.
Other pedestrian areas: Riven York stone, with granite setts around street furniture.

Signs: At entry to (Diag. 674) and exit from (Diag. 675) 20 mph zone. No intermediate signs or road markings.

Lighting: Replaced in part with period style columns and lanterns.

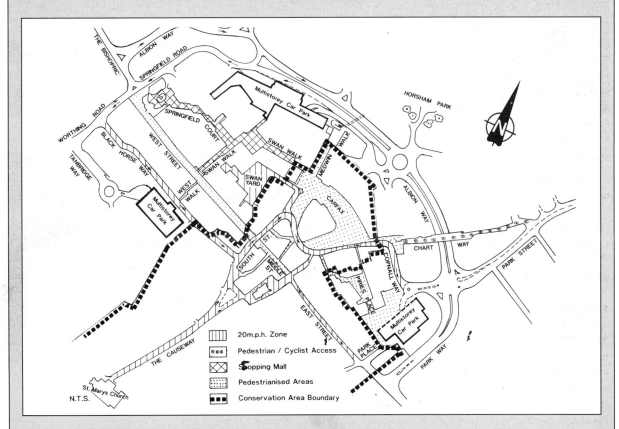

Key:
- IIII 20m.p.h. Zone
- Pedestrian / Cyclist Access
- Shopping Mall
- Pedestrianised Areas
- Conservation Area Boundary

N.T.S.

Cost: Construction costs were approximately £ 2.3 million, funded by Horsham District Council. West Sussex County Council and Horsham District Council paid their own design and site supervision staff costs.

Contact's Comments: Successfully reduced speeds and accidents. The removal of on street parking within the Carfax has reduced traffic flows. Car parks are situated adjacent to the area providing easy pedestrian access.
The scheme was winner of 1993 *Urban Street Environment* Traffic Calming Award.

2 20 MPH ZONE, VARIED MEASURES
Rushden, High Street

Location: Northamptonshire: Rushden, High Street.

Implementation Date: November 1992.

Background: An experimental pedestrianisation scheme had proved unpopular with local traders. It was agreed to replace this with a 20 mph zone in the traditional shopping area of Rushden, centred on High Street. The area consists of traditional shops and small branches of chain stores, and is bounded on both sides by the A6.

Need for Measures: To slow traffic and enhance pedestrian safety, whilst maintaining delivery access to the shops and, by improving the urban environment, stimulate increased economic activity.

Measures Installed: The measures included flat and round top road humps, pinch points, chicanes and pavement extensions.

Special Features: Red block paving was used to complement the existing brick frontages, and brick constructed planters were installed to allow floral displays. Street furniture was manufactured of cast-iron and painted black to blend in to the surroundings and maintain the character of the area.

Consultation: Public exhibition, meetings with the Chamber of Trade and District Councillors, consultation with the emergency services and bus operators.

Monitoring	Accidents (pia)	Speeds (mph)	Traffic (16 hr)
Before	11 in 3 years	24	2,541
After	3 in 18 months	17	2,080

 **Northamptonshire County Council**

Contact: **John Shortland   Tel: 01604 236713**
Authority: **Northamptonshire County Council**

Technical Data:

Location Type: Main shopping area.

Road Type and Speed Limit: Urban unclassified: 30 mph.

Scheme Type: Flat top humps, pinch points and chicanes in High Street and College Street.
Round top humps in side streets.

Special Authorisation: Temporary (12 month) 20 mph speed limit order required from DOT.

Length of Scheme in Total: Zone covers 10 hectares. 1 km of carriageway.

Dimensions: Height: 75 mm.
Width: Full road width.
Length: 6 m.
Ramp gradient: 1:12.
Distance from junctions: Varies.

Materials: Humps: Block paving for flat top humps. Blacktop for round top humps.
Kerbs: Precast concrete.
Street furniture: Cast iron bollards and litter bins. Brick planters.

Signs: As required for a 20 mph zone (Circular Roads 4/90).

Lighting: The existing street lighting was retained.

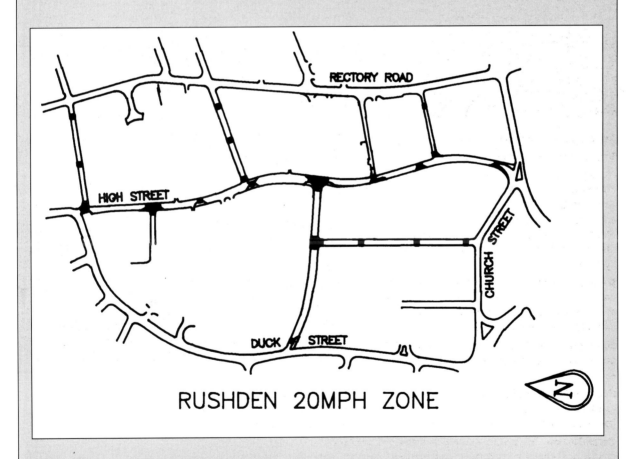

RUSHDEN 20MPH ZONE

Costs: Northamptonshire County Council £217,000.
East Northants District Council £13,000.
Total £230,000.

Contact's Comments: The scheme has been successful in reducing accidents by nearly a half,
and in reducing speeds to such an extent that the temporary 20 mph zone order could be made permanant.

3 PRO-PEDESTRIAN FEATURES
Borehamwood, Shenley Rd

Location: Hertfordshire: Borehamwood, Shenley Road.

Implemented: Experimental April 1990. Permanent 1992 - 1994.

Background: Shenley Road is an east-west route between Barnet and Watford which carries through industrial traffic, and is the main shopping centre for a population of 30,000. The trading position of the shopping street poor, with many competitor towns within 10 miles. Potential economic decline of the town gave it high priority in Hertfordshire's Town Centre Enhancement programme.

Need for Measures: Serious congestion. Illegal parking on carriageway. Some pedestrian/vehicle accidents. No suitable bypass route.

Measures Installed: Low central reserve to create dual 3.6 metre wide carriageways, horizontal deviations, mini roundabouts and road humps.

Special Features: Flat top road humps on low speed dual carriageway, with parking in marked bays.

Consultation: Public exhibitions and leaflets to all premises in September 1988, March 1990, December 1991 and May 1993. Town Centre Panel of all interested parties set up, meeting monthly since February 1990. Visits to all frontagers by officers and contractor at all appropriate stages of the scheme.

Monitoring	Accidents (pia)	Speeds (mph)	Traffic 16 hrs
Before	15	26	18,000
After	8	20	16,500

Contact: Lilian Goldberg Tel: 01992 556057

Authority: Hertfordshire County Council
Hertsmere Borough Council

Hertfordshire
COUNTY COUNCIL
Transportation

Technical Data:

Location Type: Town centre shopping street carrying through traffic.

Road Type and Speed Limit: Formerly principal road A5135, now B5378. 30 mph.

Scheme Type: Narrowing to dual one lane. Flat topped road humps.

Length of Scheme in Total: 800 m.

Dimensions:
Height: 100 mm.
Width: Each carriageway 3.75 m, plus variable central reserve.
Length: Plateaus 6 m.
Ramp gradient: 1:15 and 1:20.
Distance from junctions: Varies.

Materials:
Plateau and ramps: Red concrete blocks.
Kerbs: Natural concrete.

Signs: Diag. 562 with sub-plate 'Pedestrians Crossing on Humps' at each hump.

Lighting: Carriageway lighting - 8 m columns with large globe. Footway lighting 6 m columns with standard globe on swan neck. New lighting scheme allows extra lighting for road humps and pedestrians.

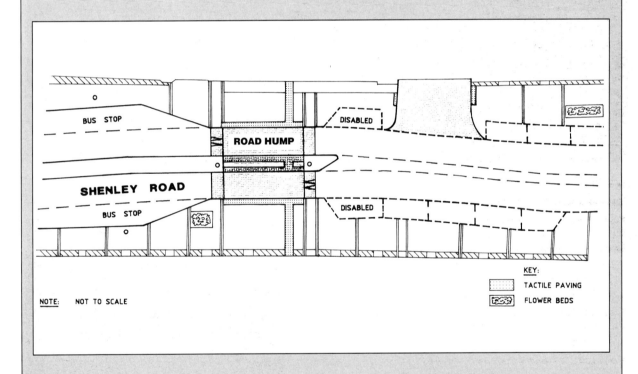

KEY:
▒ TACTILE PAVING
▒ FLOWER BEDS

NOTE: NOT TO SCALE

Cost:
Developer contribution £1 million.
County and District Councils £100,000 each.
Total £1.2 million.

Contact's Comments: The town centre has been transformed from a typical vehicle dominated environment to one which is very pedestrian friendly. The informal crossing points offer pedestrians easy and quick crossing. Vehicle flow has improved. The scheme was a winner of the 1994 *Urban Street Environment Traffic Calming Award*.

4 NARROWING, CHICANES, CYCLE LANES
Cambridgeshire: Eaton Socon

Location: Cambridgeshire: Eaton Socon.

Implemented: May 1992.

Background: The B1428 through Eaton Socon is a former trunk road which acts as a main feeder route into St Neots town centre. The route severs two large residential areas and is subject to a 30 mph speed limit. The road width varies between 7.3 and 10 metres.

Need for Measures: Persistent high injury accident rate.

Measures Installed: Mini-roundabouts, traffic islands, road narrowings and cycle lanes.

Special Features: Cycle lanes at road narrowings.

Consultation: Town Council, local residents, bus companies and emergency services.

Monitoring	Accidents (pia)	Speeds (mph)	Traffic (12 hr)
Before	56 in 3 years	38	10,700
After	6 in 18 months	32	8,900

Cambridgeshire County Council

Contact: Bob Menzies Tel: 01223 317749

Authority: Cambridgeshire County Council

Technical Data:

Location Type: Urban, market town.

Road Type and Speed Limit: Urban classified: 30 mph.

Scheme Type: Road narrowings, mini-roundabouts, traffic islands and chicanes with cycle lanes.

Length of Scheme in Total: 2.3 km.

Dimensions: Length: Chicanes and road narrowings - varies. Islands 6 m.
Overrunnable islands 2.4 m.

        Height: Overrunnable islands 40-100 mm.

        Width: Carriageway at islands 2.8-3.0 m. Cycle lanes 1.0 m.
Carriageway at chicanes/narrowings 5.0-5.5 m.
Islands 1.2-2.0m. Overrunnable islands 0.9 m.

Materials: Chicanes: Precast concrete kerbs/conservation kerbs with asphalt infill.
Traffic islands: Precast concrete kerbs/conservation kerbs with asphalt infill.

Signs: Regulatory: Islands Diag 610. Mini- roundabouts Diags. 611.1, 1003.3 and 1003.4.
Narrowings/chicanes Diag. 561. Marker posts Diag. 1011.

    Advance: Diag. 562 with supplementary plate 'Traffic Calmed Zone'.
Cycle lanes: Diags. 625.5 and1049.

Lighting: Existing 10 m lamp columns with additional 5 m lamp columns at some features.

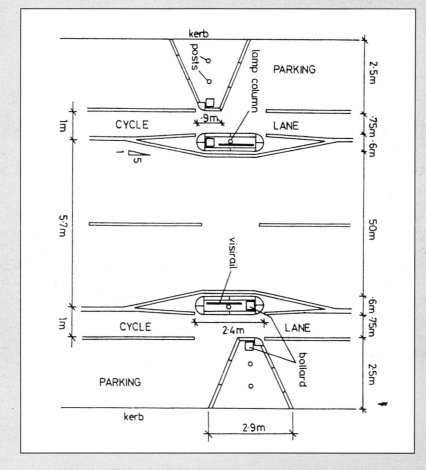

Cost: Cambridgeshire County Council Traffic Management and Safety budget £130,000

Contact's Comments: While speed reduction is not great, the measures have concentrated drivers' attention, to an extent which is reflected in the marked accident reduction.

5 NARROWING, FLAT TOP HUMP
Saltash, Fore Street

Location: Cornwall: Saltash, Fore Street.

Implemented: November 1991.

Background: Fore Street is the shopping and business centre of Saltash. It is 450 metres in length, subject to a 30 mph speed limit and has street lighting. Although chiefly a shopping centre, it is used as a rat-run by commuters short cutting across to the Tamar Bridge and into Plymouth.

Need for Measures: Speeds and rat-running, which led to a high number of accidents.

Measures installed: Five flat top road humps, carriageway narrowing, protected parking bays and environmental enhancement.

Special features: Use of tactile paving throughout, both at controlled and uncontrolled crossing points. Tree planting.

Consultation: With emergency services, Town and District Councils, Chamber of Commerce, bus companies, plus public consultation.

Monitoring	Accidents (pia)	Speeds (mph)	Traffic (12 hr)
Before	18 in 3 years	32	9,000
After	2 in 2 years	17	7,900

one and all   onen hag oll
CORNWALL
COUNTY COUNCIL

Contact: Peter Moore Tel: 01209 820611

Authority: Cornwall County Council

Technical Data: 43

Location Type: Small town.

Road Type and Speed Limit: Urban unclassified: 30 mph.

Scheme Type: Flat top humps. Carriageway narrowed to 5.5 m at humps, approx 5.8 m at other points.

Length of Scheme in Total: 450 m.

Dimensions: Height: 100 mm.
Width: 5.5 m.
Length: 3.7 m.
Ramp gradient: 1:6.

Materials: Road humps: Red clay block.
Carriageway: Black bitmac.
Footways: Small element buff slabs.
Kerbs: Re-used granite plus some new local granite.
Street furniture: Cast iron with town and county crests.

Signs: Diagrams 557.1, 547.6 and 543. Road markings: Diag. 1061. Tactile paving.

Lighting: Three additional lights provided.

Cost: Cornwall County Council £130,000
ERDF grant £100,000
Total £230,000

Contact's Comments: The cost includes the environmental works which involved carriageway reconstruction, resurfacing of all the footways, planting of semi-mature trees and shrubs and the introduction of cast iron street furniture.

6 NARROWING, FLAT TOP SPEED TABLE
Barnstaple, Tuly Street

Location: Devon: Barnstaple, Tuly Street.

Implemented: January 1989.

Background: Tuly Street was originally a narrow street providing some access to the Library, Cattle Market, car parks and some shops in the High Street. Since pedestrianisation of High Street all daytime access is to the rear via Tuly Street. A car park which caters for short stay makes Tuly Street busy with pedestrian crossover flows of up to 1,000 per hour.

Need for Measures: To slow traffic and provide easier crossing point for pedestrians.

Measures Installed: Single lane narrowing and raised platform for two-way working with no priority marked.

Special Features: Brick carriageway, bollards and other street furniture to emphasize narrowing and match adjoining development.

Consultation: District Council, developers and utilities.

Monitoring	Accidents (pia)	Speeds (mph)	Traffic (16 hr)
Before	1 in 4 years	23	1,800
After	1 in 4 years	11	4,000

DEVON
COUNTY COUNCIL

Contact: David Netherway Tel: 01271 388503
Authority: Devon County Council

Technical Data:

Location Type: Town centre.

Road Type and Speed Limit: Urban unclassified: 30 mph.

Scheme Type: Narrowing - single way working with flat top speed table.

Length of Scheme in Total: 31 m.

Dimensions::   Height: 90 mm (av).
Width: 3 m.
Length: Plateau 28 m. Ramps 1.5 m.
Ramp gradient: 1:16 (av).
Distance from junctions: 83 m.

Materials: Plateau and ramps: Brindle brick paving with two bands of blue block paviours to denote crossing location.

Kerbs: Precast concrete.

Street furniture: Cast iron bollards and fingerpost.

Signs: None.

Lighting: Ornamental double arm column and lantern.

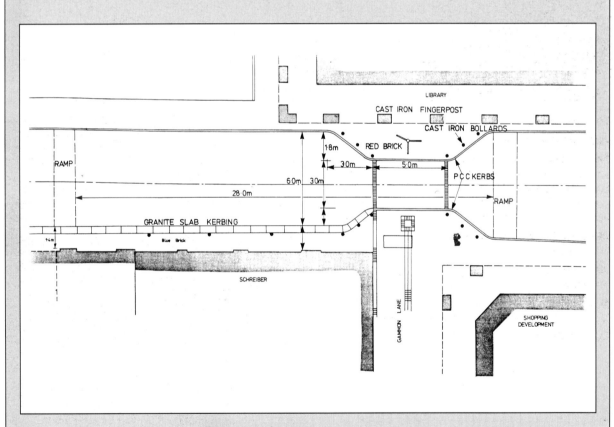

Cost: Funded by developer £3,000.

Contact's Comments: Works very well, no priority signing necessary. Traffic gives way in turn with pedestrians having about equal priority with traffic.

7 NARROWING, FLAT TOP HUMPS, PELICANS
Southend, Marine Parade

Location: Essex: Southend-on-Sea, Marine Parade.

Implemented: March 1992.

Background: Marine Parade forms part of the 'Golden Mile' of seafront in Southend. It is a two lane dual carriageway subject to 30 mph speed limit. The road is bordered by amusement arcades, funfairs and the usual seafront activities, and because it had a roundabout at each end, suffered from 'cruising' problems. Drivers used to come from up to 100 miles away to circuit the roundabouts.

Need for Measures: Before any measures were taken, there were 28 personal injury accidents per year. Alterations to the western roundabout the year before the traffic calming produced no reduction. The measures were also designed to assist the police in controlling cruising problems.

Measures Installed: One roundabout altered to form 'right turn only' lane , thus preventing u-turns around 'the circuit'. Two lane dual carriageway road reduced to single lane in each direction. Six flat top road humps installed, three in each direction. Subsequently two pelican crossings added to control pedestrian flow over humps.

Consultations: Statutory bodies consulted. An attitude study was carried out post implementation.

Monitoring	Accidents (pia)	Speeds (mph)	Traffic Flow
Before	28 in 1 year	28	-
After	9 in 1 year	26	-

Contact: Nicola Foster   Tel: 01245 437113
Authority: Essex County Council

Essex County Council
Highways

Technical Data

Location Type: Seafront main road in town with seasonal leisure activity.

Road Type and Speed Limit: Urban distributor 'B' road, two-lane dual carriageway: 30 mph.

Type of Scheme: Flat top humps narrowing to single lane.

Length of Scheme in Total: 0.5 km.

Dimensions: Width: 4.5 m.
Length: 8 m flat top.
Ramp gradient: 1:12.

Materials: Block topped with hot rolled asphalt ramps between kerbs, where pedestrians encouraged to cross.
Hot rolled asphalt topped where designed for vehicles only.

Signs: Diags. 516, 615, 811 and 1061.

Lighting: As required.

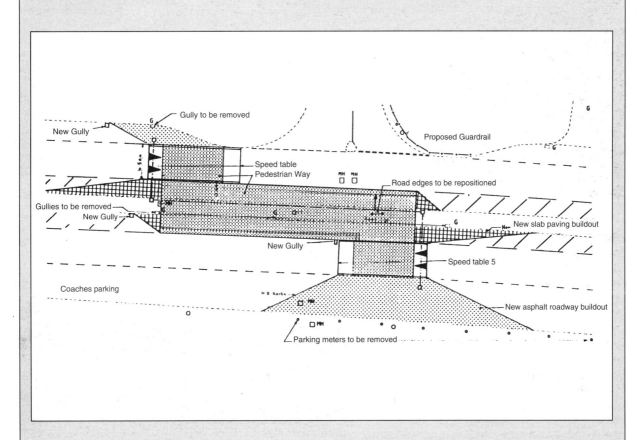

Costs: Civils £95,000.
Associated works £3,000.
Pelicans £76,000.
Monitoring (including videos and surveys) £ 7,000.
Total £181,000. Funded from Essex County Council Capital budget.

Contact's Comments: It is interesting to note that six of the nine injury accidents after installation were rear-end shunts and all occurred between 21.15 and 23.49 hours. With hindsight, traders, frontagers and residents should have been consulted in advance. Very effective in reducing pedestrian injury accidents.

8 ROUND TOP HUMPS , DUAL CARRIAGEWAY
Gillingham, Twydall Green

Location: Kent: Gillingham, Twydall Green.

Implemented: October 1988.

Background: The Twydall Estate shopping and community centre is bisected by a short length of dual carriageway. It is within the 30 mph limit and is a bus route.

Need for Measures: Pedestrian accidents, including one fatal in the three years prior to 1988. Contributory factors were unreasonably high speeds and pedestrians crossing between vehicles parked on both sides including up against the central reservation.

Measures Installed: A series of three DOT-style round top humps, with blipped kerbs sited at the humps to create parking bays, and guard railing on the central reserve. Gaps in the railing at each hump allow pedestrians to cross. Some lighting columns were also re-sited.

Special Features: Profile of humps was flattened to allow pedestrians to walk on them.

Consultation: With the District Council and emergency services. There was no formal public consultation as this was an early scheme before KCC procedures had been finalised.

Monitoring	Accidents (pia)	Speeds (mph)	Traffic Flow (12 hr)
Before	4 in 3 years	not measured	5,000
After	2 in 3 years	not measured	5,000

Contact: Ray Dines   Tel: 01622 696993
Authority: Kent County Council

Kent County Council
HIGHWAYS &
TRANSPORTATION

Technical Data: 49

Location Type: Urban shopping and community centre.

Road Type and Speed Limit: Urban unclassified: 30 mph.

Type of Scheme: Round top humps and sheltered parking on a dual carriageway road.

Length of Scheme in Total: 130 m.

Dimensions: Height: 100 mm - flattened profile.
Width: Full width.
Length: 3.75 m.
Ramp gradient: N/A.
Distance from junctions: Within 40 m.

Materials: Road humps: Hot rolled asphalt.

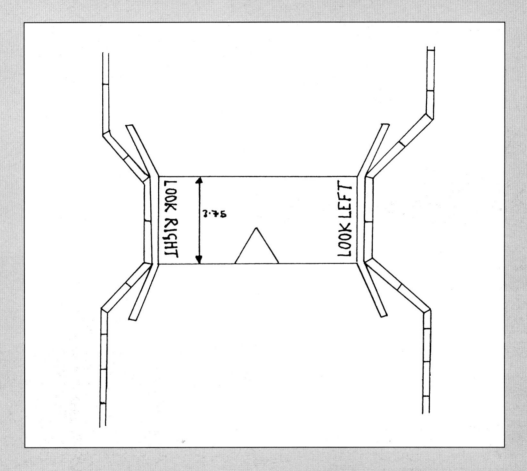

Cost: £18,000.

Contact's Comments: Bus company has used the scheme without comment since its implementation.

9 FLAT TOP HUMPS
Rochester, Vines Lane

Location: Kent: Rochester, Vines Lane.

Implemented: May 1990.

Background: Vines Lane is on the periphery of the historic centre of Rochester. It is some 200 metres in length, subject to a 30 mph speed limit and has lighting. It is used as a commuter cut-through and is bounded by a municipal park and school playing fields.

Need for Measures: Speeds which led to child pedestrian accidents and rat-running.

Measures Installed: Two single-way working ramps within 50 m of junctions, each giving priority from main road.

Special Features: Consideration given to scheme blending in with historic setting.

Consultation: With the local school and the police. No residents adjacent to this road, no bus route.

Monitoring	Accidents (pia)	Speeds (mph)	Traffic
Before	6 in 3 years	35	8,000
After	0 in 3 years	25	4,000

Contact: Theresa Trussell   Tel: 01622 696876
Authority: Kent County Council

Kent County Council
HIGHWAYS & TRANSPORTATION

Technical Data:

Location Type: City centre conservation area.

Road Type and Speed Limit: Urban unclassified: 30 mph.

Scheme Type: Flat top humps.

Length of Scheme in Total: 235 m.

Dimensions: Height: 100 mm.
Width: 3 m.
Length: Plateau 10 m. Ramp 1 m.
Ramp gradient: 1:10.
Distance from junctions: 50 m.

Materials: Plateaus: Red block paving with 2 m central band of grey block paving to denote crossing point.
Ramps: Recycled granite setts.
Kerbs: Recycled granite.
Street furniture: Bollards - cast iron with city crest.

Signs: Regulatory sign: Diag. 811. 'Road Humps' sign: Diag. 557.

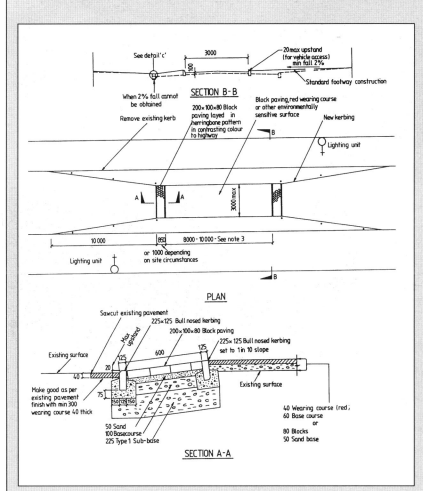

Lighting: Existing period style lamp columns resited and up-graded.

Cost: Kent County Council Small Improvement budget £10,500.

Rochester upon Medway City Council for environmental enhancements £5,500.

Kings School donation £2,500.

Total £18,500.

Contact's Comments: Recent study suggests overall drop in accidents in locality, with no evidence of accident migration.

10 NARROWING, HUMPED CHICANE
Ryde, North Walk

Location: Isle of Wight: Ryde, North Walk.

Implemented: April 1992.

Background: North Walk is a section of urban highway that runs adjacent to Appley Beach, from Ryde Esplanade in the west to Appley Park in the east. Appley Park is a large wooded recreational area that contains a wide range of leisure activities, including a coastal walk.
There is also a 119 space car park and an inshore lifeboat station. The combination of all of these activities created a significant traffic hazard at peak times, particularly to pedestrians.

Need for Measures: The speed of traffic exiting and entering this recreational area caused a potential hazard to both pedestrians and other road users.

Measures Installed: Two single way narrowings and a central humped chicane.

Special Features: None.

Consultation: Carried out by the Borough Council.

Monitoring	Accidents (pia)	Speeds (mph)	Traffic (12 hr)
Before	0	29	850
After	0	18	815

Contact: P Taylor   Tel: 01983 823763
Authority: Isle of Wight County Council

Technical Data:

Location Type: Urban, coastal recreation area.

Road Type and Speed Limit: Urban unclassified: 30mph.

Scheme Type: Narrowing - single way working with central humped chicane.

Length of Scheme in Total: 30 m.

Dimensions: Height: Humped chicane 100 mm.
                  Width: 2 m.
                  Length: 9 m.
                  Distance from junctions: 10 m.

Materials: Kerbs: Hydraulically pressed pre-cast concrete.
                  Street furniture: Priority signposts.

Signs: Diags. 615, 602 and 811.

Lighting: STD amenity street lighting, replacing poor sub-standard lighting.

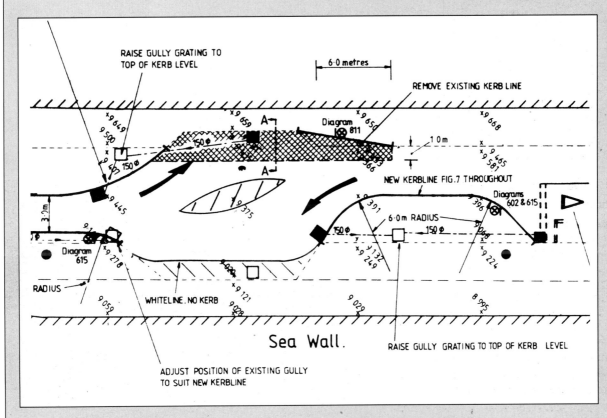

Cost: £14,500.

Contact's Comments: None supplied.

11 ADJACENT RAISED MINI-ROUNDABOUTS
Leamington Spa, The Parade

Location: Warwickshire: Leamington Spa, The Parade.

Implemented: July 1992.

Background: This road carries traffic through the middle of the town centre, bisecting large national retail shops. Traffic type includes buses, delivery vehicles and the emergency services.

Need for Measures: High injury accident record, particularly involving pedal cyclists.

Measures Installed: Two speed tables with mini-roundabouts and associated signs and road markings.

Special Features: Clay paviours used initially but replaced with asphalt because of low polished paviour value and skidding problems.

Consultation: No residents, otherwise full legal consultation. No objections were received from the emergency services or bus operators.

Monitoring	Accidents (pia)	Speeds (mph)	Traffic Flow (16 hr)
Before	9 in 3 years	not known	15,600
After	9 in 22 months	20	17,500

Warwickshire County Council

Contact: Paul Bellotti Tel: 01926 412179
Authority: Warwickshire County Council

Technical Data:

Location Type: Town centre.

Road Type and Speed Limit: Urban classified: 30 mph.

Scheme Type: Two raised mini-roundabouts.

Length of Scheme in Total: 50 m.

Dimensions: Height: 100 mm.
Width: 15 m.
Length: 20 m.
Ramp gradient: 1:20.
Distance from junctions: On junction.

Materials: Plateau and ramps: Asphalt.
Kerbs: Concrete.
Street furniture: Black cast iron bollards to define footway.

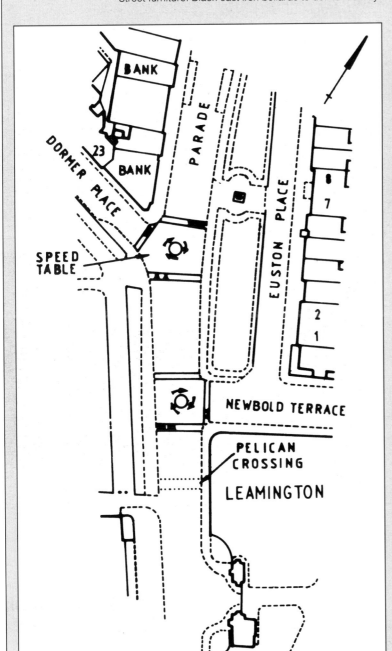

Signs: Advanced and local warning signs together with 'mini-roundabout' sign.

Lighting: High pressure sodium (as existing).

Cost: Warwickshire County Council Local Safety Scheme budget £35,000.

Contact's Comments: During a nearby road closure for bridge works (effective 14/09/92 to 4/12/93) traffic flows became much higher through the junctions, and five of the nine accidents occurred during that time. Changes to the entry ramp slopes are being considered as the 1:20 gradient may be too relaxed.

12 MINI-ROUNDABOUT, FLAT TOP HUMPS
Rugby, Ashlawn Road

Location: Warwickshire: Rugby, Ashlawn Road.

Implemented: January 1992.

Background: This road carries a mixture of both local and inter-urban traffic avoiding the town centre. To one side of the road is a large school and to the other there is housing.

Need for Measures: High accident record. Very heightened local concern was felt and pressure exerted for traffic calming.

Measures Installed: One junction improvement, one raised mini-roundabout and eight flat top humps.

Special Features: Raised mini-roundabout constructed in blockwork on a 'B' road.

Consultation: Full legal consultation including leaflet drop, public exhibition for residents. No objections were received from the emergency services.

Monitoring	Accidents (pia)	Speeds (mph)	Traffic (16 hr)
Before	6 in 3 years	50	7,100
After	0 in 25 months	25	5,400

Warwickshire County Council

Contact: Phil Cook Tel: 01926 412345
Authority: Warwickshire County Council

Technical Data:

Location Type: Edge of town centre.

Road Type and Speed Limit: Urban classified: 30 mph.

Scheme Type: Raised mini-roundabout and flat top humps.

Length of Scheme in Total: 550 m.

Dimensions: Height: 75 mm (max).
Width: 6.5 - 7.0 m.
Length: 3.7 m.
Ramp gradient: 1 in 15.
Distance from junctions: Variable.

Materials: Plateau and ramps: Block paving.
Kerbs: Concrete.
Street Furniture: Steel posts incorporating retro-reflective red/white material.

Signs: Advanced and local warning signs together with 'mini-roundabout' sign.

Lighting: Low pressure sodium (as existing).

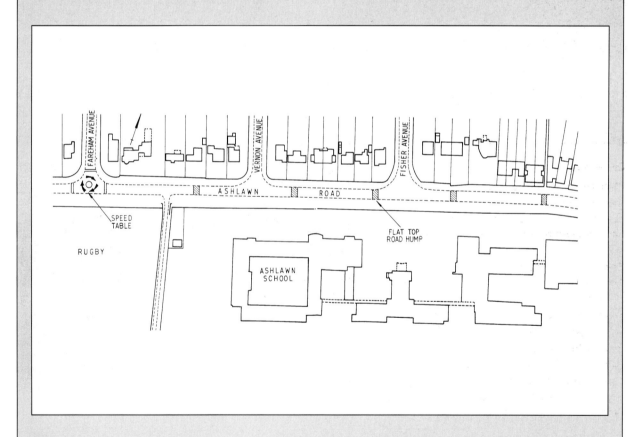

Cost: Warwickshire County Council Local Safety Scheme budget £69,000.

Contact's Comments: Ramp slopes modified from 1:10 to 1:15 after complaints from public about the severity of the measures.

13 BYPASS DEMONSTRATION PROJECT
Petersfield, Dragon Street

Location: Hampshire: Petersfield, Dragon Street.

Implemented: August 1993.

Background: Petersfield was one of six towns across the country to be selected by the Department of Transport for inclusion in a national project on how to maximise the environmental and safety benefits to towns relieved of trunk road traffic by new bypasses. Over a three year period, major re-modelling and environmental enhancement of the old A3 route through Petersfield is being undertaken.

Need for Measures: To restore a 'sense of place' to Petersfield, to reduce the dominance of traffic, to provide a safe and pleasant environment for residents, shoppers and visitors, and to preserve and enhance local business viability.

Measures Installed: Phase I of the project was to design improvements for Dragon Street, formerly part of the heavily trafficked through route. The scheme consists of a substantial narrowing of the road and widening of the pavements.

Special Features: See materials.

Consultation: East Hampshire District Council, Petersfield Town Council, Petersfield Chamber of Trade, local shopkeepers, Residents Groups and the Petersfield Society conducted through a local Forum chaired by the Town Mayor. Public meetings and articles in the press are being used to inform local residents, and to receive feedback.

Monitoring: Changes in traffic patterns, vehicle speeds, journey times, pedestrian and cyclist movements, noise and air quality are being monitored. Similar studies are being undertaken in the other five towns participating in the demonstration project. The data collected will help construct a comprehensive assessment of changes in the six towns to guide other local authorities.

Contact for further details: Graham Carter Tel: 01962 846127
Authority: Hampshire County Council

Hampshire
COUNTY COUNCIL

Technical Data:

Location Type: Market town. Conservation area.

Road Type and Speed Limit: Formerly A3 trunk road. Now B2070 secondary road. 30 mph.

Scheme Type: Narrowed two-way carriageway permitting widened footways incorporating parking bays.

Length of Scheme in Total: 400 m.

Dimensions: Carriageway width reduced from 10 m to 6.75 m.

Materials:
Junctions: Bleijko blocks in autumn colour mix.
Buff blocks to denote pedestrian crossing points and channels.
Granite rumble strips at boundaries.

Footways: 3 sizes of paving slab. Blockwork drainage runs. Conservation kerbing.

Parking bays: Dark blocks.

Street furniture: Locally sourced wrought iron grills and guards. Back cycle stands.
Wooden slatted benches. Bus shelters.

Tree planting: Semi-mature specimens including ornamental pears, planes and alder
to provide a sense of enclosure.

Signs: Cast iron direction signs including black/gold pedestrian finger posts. No road markings.

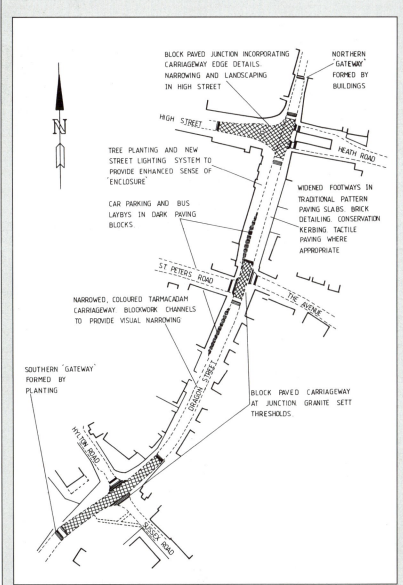

Lighting: 6 m high black columns specially commissioned from Urbis which contribute greatly to the smaller scale street scene.

Costs (Phase 1):

DOT Bypass Demonstration Project Grant £193,500.

Supplementary Credit Approval £193,500.

De-trunking Maintenance Grant £93,000.

Total: £480,000.

Contact's Comments: The scheme has been well received by local people. For instance a mural celebrating the completion of the bypass has been sited in Dragon Street and local properties are now being refurbished. There are also reports of increased trade in town centre shops, and the results of the data and attitude surveys are eagerly awaited. The scheme was a finalist for 1994 *Urban Street Environment* Traffic Calming Award.

Rural

Case Studies 14 – 38

14 VISP, PINCH POINTS, BUILD-OUT
Somerset: Stratton-on-the-Fosse

Location: Somerset: Stratton-on-the-Fosse village.

Implemented: May 1993.

Background: The A367 Fosse Way is a principal county route which runs through the village of Stratton-on-the-Fosse. The village is situated between Shepton Mallet and Radstock.

Need for Measures: To reduce traffic speed and provide safer conditions for pedestrians.

Measures Installed: Pinch-points and a single-side build-out, with priority control signing.

Special Features: Exposed aggregate concrete planter tubs and hazard marker bollards to emphasise pinch-point features.

Consultation: Parish Council, local residents, local County Councillor, Police, emergency services, bus companies and utilities.

Monitoring	Accidents (pia)	Speeds (85%ile mph)	Traffic (16 hr)
Before	9 in 5 years	47	4,000
After	1 in 1 year	35	5,000

Contact: Stephen Sully Tel: 01823 255630
Authority: Somerset County Council

Somerset County Council

RURAL

Technical Data:

Location type: Village.

Road Type & Speed Limit: Principal county route A367: 30mph.

Scheme Type: Pinch points - single lane working with alternating priority control signing at features.

Length of Scheme in Total: 1 km.

Dimensions: Reduced carriageway width at features: 3.2 m.
Length of features: 9 m (av).

Materials: Pinch points: Precast concrete kerbing with bitmac infill and 'Give Way' markings on carriageway.

Street Furniture: Plastic hazard marker bollards and concrete planter tubs.

Signs: Priority control with chevrons, externally illuminated.

Lighting: Measures sited adjacent to existing street lights.

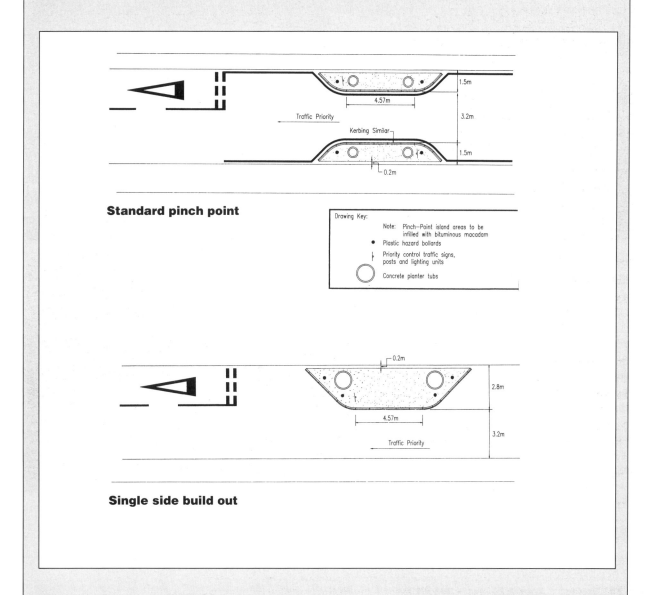

Standard pinch point

Drawing Key:

Note: Pinch–Point island areas to be infilled with bituminous macadam
• Plastic hazard bollards
Priority control traffic signs, posts and lighting units
◯ Concrete planter tubs

Single side build out

Cost: Somerset County Council £50,000.

Contact's Comments: The scheme, which is one of the 'VISP' village speeding demonstration projects, has proven extremely effective in reducing traffic speeds and creating safer conditions for pedestrians. Measures rely on a degree of opposing traffic to maintain speed reduction.

15 VISP, GATEWAYS
Devon: Halberton

Location: Devon: Halberton.

Implemented: June 1992.

Background: Halberton village lies on a route which is now by-passed by the North Devon Link Road, but still suffers from high speed traffic, mainly locally generated.

Need for Measures: To reduce traffic speed and improve road safety.

Measures Installed: Gateways were provided at each end of the village, located near the boundary of the 30 mph limit. These consisted of central traffic islands with 'keep left' bollards and high level beacons, and a red coloured asphalt surface with '30 mph' roundels. White hatching at the side of the road, with reflector posts, further reinforces the narrowing effect.

Special Features: Village nameplates are also located at the gateway.

Consultation: Local members, the Parish Council and the emergency services

Monitoring	Accidents (pia)	Speeds (mph)	Traffic (16 hr)
Before	2 in 3 years	44	2,000
After	3 in 2 years	36	2,000

DEVON
COUNTY COUNCIL

Contact: Richard Oldfield   Tel: 01392 383800
Authority: Devon County Council

Technical Data:

Location Type: Village approach.

Road Type and Speed Limit: Class C: 30 mph.

Scheme Type: Gateways.

Dimensions: Widths: Lanes 3 m. Central island 1.2 m.

Materials: Coloured asphalt: red.

                 Calcined flint surface dressing - white.

Signs: Temporary 'New traffic island ahead' signs.

Lighting: New lighting at eastern end of village, plus
                 high level illuminated beacons on traffic islands

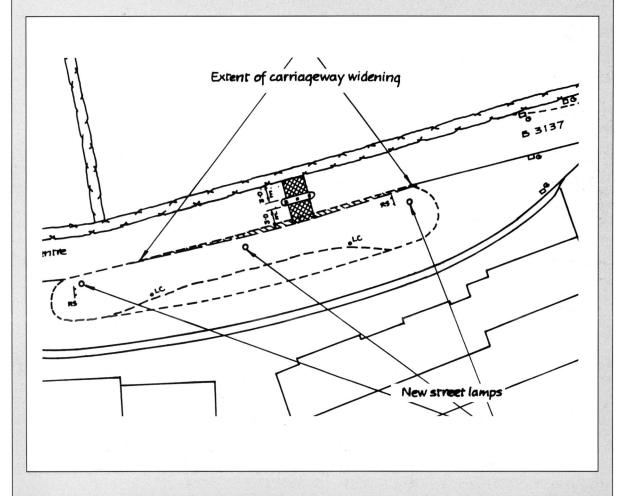

Cost: Devon County Council £9,000.

Contact's Comments: The scheme has been successful in reducing speeds at the gateway sites,
but reductions in the village itself have been very slight. Accident rates have not shown a reduction,
but with the small numbers involved this is not considered to be statistically significant.

16 VISP, GATEWAYS
Kent: Matfield

Location: Kent: Matfield village.

Implemented: September 1992.

Background: Matfield village lies on the B2160 about 8 km east of Tunbridge Wells and has an estimated population of 1,100. Traffic flows are moderate, with a six day two-way mean (before the measures) of 6,900 vehicles per day through the village centre. There is a 30 mph speed limit through the village centre. There is no street lighting.

Need for Measures: Speeds, which led to accidents involving pedestrians and parked cars.

Measures Installed: 'Gateway' signing. This VISP scheme combines, for the first time in the UK, a village nameplate with speed limit regulatory signs. These were located on the major road approaches.

Special Features: Innovative signs which required special authorisation from the Department of Transport.

Consultation: With the Parish Council and the police.

Technical Data:

Location Type: Village approach.

Road Type and Speed Limit: Rural 'B' road: 30 mph.

Length of Scheme in Total: 1 km.

Signs: At both the north and south gateways:

Nearside: Enlarged village name plate sign (Diag 161.5 at 735 mm) incorporating '30' roundel and the message 'Please Drive Carefully through the Village'.

Offside: Plate (Diag 1075 at 875 mm) displaying a '30' roundel with the village name underneath.

Both signs backed with 'derestriction' sign. The larger sign also shows the message 'Thank You for Driving Carefully'.

Complementary signing of a similar style was installed on the eastern and western minor road approaches.

Cost: £2,200.

Contact's Comments: All accidents slight.

Monitoring	Accidents (pia)	Speeds (mph)	Traffic
Before	7 in 3 years	40	6,900
After	3 in 18 months	38	6,900

Contact: Theresa Trussell Tel: 01622 696876
Authority: Kent County Council

17 SPEED CAMERA
Oxfordshire : Nuneham Courtenay

Location: Oxfordshire: Nuneham Courtenay.

Implemented: July 1993.

Background: Nuneham Courtenay is a small 'one street' village, with development extending approximately 0.5 km along both sides of the busy A4074 (formerly A423(T)) road. The road is subject to a 40mph limit in the village.

Need for Measures: There are two accident clusters in the village. At the south end accidents were recorded at a major/minor junction just south of the 40 mph limit; a right turn lane was provided in 1991. At the north end of the village a number of accidents involving turning movements to and from a restaurant were associated with restricted visibility. Speed measurement confirmed a significant level of abuse of the speed limit.

Measures Installed: One pole mounted speed camera and, later, gateways.

Special Features: The village is a Conservation Area, and this places a constraint on the choice of appropriate measures. Even the provision of a speed camera resulted in some concern from the District Council, although the measure was welcomed by the Parish Council.

Consultation: With police only.

Technical Data:

Location Type: Village.

Road Type and Speed Limit: Principal road: 40 mph.

Scheme Type: Speed camera.

Length of Scheme in Total: Single point measure.

Dimensions: Overall height of camera post approximately 3.5 m.

Materials: Gatso speed camera installation, with dummy flash system operating when camera not present.

Signs: 'Speed Camera' warning sign provided for northbound traffic.

Lighting: None.

Cost:
Fixed equipment (including pro rata share of speed camera, with 1 camera serving 8 poles) £12,000.
Annual maintenance and power costs £500.
Annual police support costs approximately £4,500.

The above costs do not take into account the revenue derived from the fines paid by offending drivers. The annual income from a typical camera site is approximately £10,000.

Contact's Comments: The installation has been well received by residents and has proved effective in reducing speeds in both directions (although enforcement only carried out for northbound traffic). Gateway treatments introduced after the speed camera was installed did not enhance the speed reduction. A warning sign for the speed camera was provided for north bound traffic and was judged to enhance the effectiveness of the scheme.

Monitoring	Accidents (pia)	Speeds (mph)	Traffic (24 hr)
Before	4 in 3 years	47	17,500
After	1 in 11 months (vehicle defect)	41	17,500

Contact: Ann Mortlock   Tel: 01865 810400

Authority: Oxfordshire County Council

18 40 MPH ZONE, RURAL ROAD HIERARCHY
Hampshire: New Forest

Location: Hampshire: New Forest.

Implemented: 1990 to date.

Background: Established in 1079 by William I as a royal hunting ground, the New Forest Heritage Area covers 580 sq km, of which 198 sq km is unenclosed common grazing land. As well as trunk and principal roads, it contains approximately 400 km of Class 2, Class 3 and unclassified roads.

Need for Measures: Increasing and unrestricted traffic flows, often travelling at high speeds, were causing the death of about 150 stock animals and 70 deer per year. Over-running and damage to the verge was occurring and repairs or reconstruction of the haunch were producing increase in carriageway width.

Measures Installed: An area wide 40 mph zone, a 3-tier road hierarchy with direction signing designed to keep traffic on the highest category of road, and close liaison with the police.

Special Features: To minimise clutter within the Forest road markings rather than post mounted speed limit signs have been used, 'gateways' and chicanes have been constructed, plus 'reminder' and (regularly changed) 'slogan' signs.

Consultation: Hampshire County Council, New Forest District Council, the New Forest Verderers, the Forestry Commission and the Nature Conservancy Council, the police, the public and parish councils.

Monitoring	Accidents	Speeds
Before		80% over 44mph
After	40% reduction to humans	80% under 40mph
	35% reduction to animals	

Hampshire
COUNTY COUNCIL

Contact: Jim Soutar Tel 01962 846981

Authority: Hampshire County Council

Technical Data:

Location Type: Rural. Popular tourist destination and recreation area.

Road Type and Speed Limit: Rural 'B', 'C' and unclassified roads: 40 mph.

Scheme Type: Rural highway strategy incorporating speed limit with supporting features.

Length of Scheme in Total: Approximately 300 km.

Dimensions: N/A.

Materials: Locally sourced rough hewn timber posts for zone entry, slogan and repeater signs.

Signs: Unique New Forest style zone entry, slogan and repeater signs authorised by DOT.

Lighting: None.

Cost: Pre-1992 from Highway Maintenance budget.
1992/93 £200,000.
1993/94 £100,000.
1994/95 provisional £100,000.
Total £400,000 (plus pre-1992 costs).

Contact's Comments: The programme has been supported widely, but the production of the Strategy
was only achieved by cooperation among the numerous bodies that have an interest in the Forest. Close liaison
has been maintained during the implementation phase. The success has been greatly helped by the
personal commitment of the previous Area Surveyor, Roger Penny, who retired in 1994.

19 NARROWING, ROAD HUMPS
Cambridgeshire: Gamlingay

Location: Cambridgeshire: Gamlingay village.

Implemented: September 1990.

Background: The B1040 through Gamlingay has a generally straight alignment with a road width varying between 5 and 7 metres. The road is subject to a 30 mph speed limit and is used as a rat-run between the A1 and A45 trunk roads.

Need for Measures: High vehicle speeds and accident problems.

Measures Installed: Eight round top and two flat top road humps, two priority road narrowings and three traffic islands.

Special Features: None.

Consultation: Parish Council, bus companies and emergency services.

Monitoring	Accidents	Speeds (mph)	Traffic (12 hr)
Before	8 in 3 years	37	1,700
After	1 in 3 years	27	1,500

Cambridgeshire County Council

Contact: Richard Preston   Tel: 01223 317727

Authority: Cambridgeshire County Council

Technical Data:

Location Type: Village.

Road Type and Speed Limit: Urban classified: 30 mph.

Scheme Type: Road humps with road narrowings and traffic islands.

Length of Scheme in Total: 1.2 km.

Dimensions: Height: Round top humps 50-100 mm. Flat top humps 100 mm.

Length: Round top humps 3.7 m. Flat top 5.5 m. Islands 3.65 m.
Ramps: Flat top 1.5 m.

Width: Carriageway at islands and road narrowings 2.8-3.0 m.

Materials: Plateau: Grey concrete blockwork.
Ramps: Medium temperature asphalt.
Round top humps: Medium temperature asphalt.
Traffic islands: Asphalt infill.
Kerbs: Precast concrete.
Street furniture: Cast iron bollards.

Signs: Road humps: Diags. 557.1, 1060.1 and 1061.
Road narrowings: Diags. 503, 602, 615 and 811.
Advance: Diag. 562.

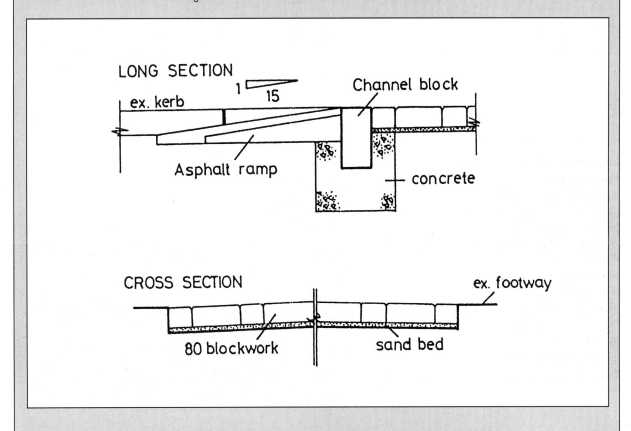

Lighting: Existing 10 m lamp columns complemented by additional lamps at some features.

Cost: £32,000 Cambridgeshire County Council Traffic Management and Safety budget.

Contact's Comments: Highly successful scheme which has addressed both the speeding and accident problem. Before/after perception surveys indicated significant change in the public perception of dangers along the route.

20 NARROWINGS, CHICANES, REFUGES
Cambridgeshire: Fen Ditton

Location: Cambridgeshire: Fen Ditton village.

Implemented: June 1992.

Background: The B1047 through Fen Ditton acts as a main feeder road into the north east of Cambridge. The generally straight route which passes a primary school has sporadic frontage development and is subject to a 30 mph speed limit.

Need for Measures: High speeds resulting in a high injury accident rate including two fatal accidents.

Measures Installed: Eight chicane effect road narrowings and seven traffic islands.

Special Features: Dual use cycle tracks to bypass road narrowings.

Consultation: Parish Council, local residents and emergency services.

Monitoring	Accidents (pia)	Speeds (mph)	Traffic (12 hr)
Before	17 in 3 years	47	10,600
After	4 in 18 months	41	11,300

Cambridgeshire County Council

Contact: Richard Preston Tel: 01223 317727
Authority: Cambridgeshire County Council

Technical Data:

Location Type: Village.

Road Type and Speed Limit: Urban classified: 30 mph.

Scheme Type: Road narrowings, chicanes and traffic islands.

Length of Scheme in Total: 1.2 km.

Dimensions: Length: Chicanes 10 m. Islands 2.7-4.5 m.
Ramps 1.5 m.

Width: Carriageway at islands 2.8-3.0 m. Islands 0.9-1.5 m.
Carriageway at chicaned narrowings 6.0 m.

Materials: Chicanes: Precast concrete kerbs with asphalt or grass infill.
Traffic islands: Precast concrete kerbs with blockwork infill.

Signs: Regulatory: Diag. 610 at islands.
Road narrowings: Diag. 561 marker posts.
Advance: Diag. 562 with supplementary plate 'Speed Reducing Measures Ahead'.
Dual use track: Diag. 956.

Lighting: Existing 5 metre lamp columns complemented by additional lamps at some features.

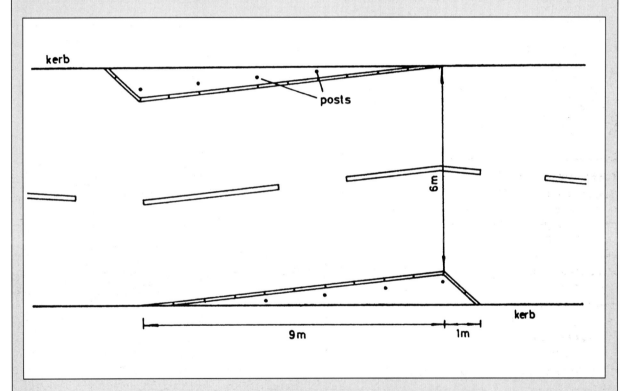

Cost: Cambridgeshire County Council Traffic Management and Safety budget £48,000.
Local community contribution £6,000.
Total £54,000.

Contact's Comments: While the scheme has achieved a significant cut in accidents, speed reduction is disappointing.
Reducing the six metre standard road width at the road narrowing may achieve further speed reductions.

21 NARROWING, SPLITTER ISLAND
Devon: Newton Tracey

Location: Devon: Newton Tracey village.

Implemented: October 1992.

Background: The village of Newton Tracey lies on a 'B' class road which is the main route between Barnstaple and Torrington. It is a bus route and also carries a high proportion of Heavy Goods Vehicles. A straight downhill approach into the village was causing traffic to speed up as they entered the narrow road through the built-up area.

Need for Measures: To reduce speeds which were causing accidents and damage to property.

Measures Installed: Central 'splitter' island with horizontal deviation, contrasting surface strips, street lighting, extra signing and landscaping to break up open area.

Special Features: 'Gateway' scheme emphasising entry into village by visual narrowing.

Consultation: Parish Council and utilities.

Monitoring	Accidents (pia)	Speeds (mph)	Traffic (16 hr)
Before	1 in 3 yrs	38 mph	2,900
After	0 in 1 yr	35	2,900

Contact: David Netherway   Tel: 01271 388503
Authority: Devon County Council

Technical Data:

Location Type: Village approach.

Road Type and Speed Limit: Rural 'B' class: 30 mph.

Scheme Type: Narrowing of approach with central 'splitter' island with extra signing, lighting and contrasting strip to form gateway.

Length of Scheme in Total: 130 m.

Dimensions: Widths: Approach 5.5 m. Lanes 3.5 m.
Lengths: Narrowing 75 m. Island 22 m. Contrasting strip 19 m.

Materials: Contrasting strip: Surface dressed with 14 mm aggregate.
Kerbs: Countryside kerbs to carriageway edge. Precast concrete full batter to island.
Island: Simulated granite setts surface with planting in centre.

Signs: Regulatory signs: '30 mph limit'.
Others: Village name boards both sides.

Road Markings: Diags. 1024,1040 and 1014.

Lighting: Double armed steel column in line with signing and illuminated 'keep left' bollards.

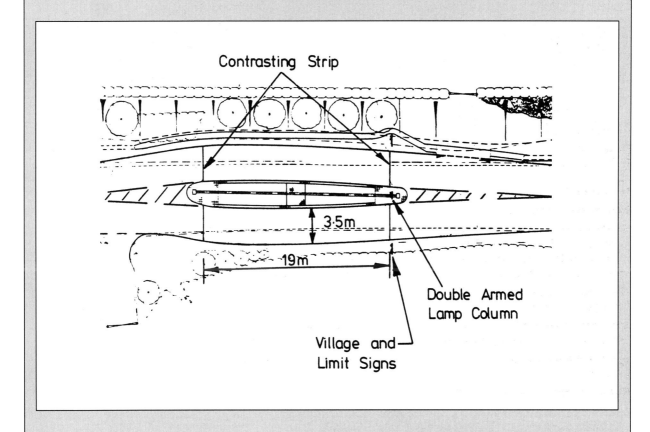

Cost: Devon County Council £20,000.

Contact's Comments: Coarse surface of contrasting strips initially gave rise to noise nuisance, subsequently surfaced over to resolve the problem. Scheme blends in well environmentally.

22 VILLAGE ENHANCEMENT, VARIED MEASURES
Hertfordshire: Buntingford

Location: Hertfordshire: Buntingford, High Street.

Implemented: August - December 1989.

Background: With the opening of a bypass in 1987 the High Street had a large drop in vehicle volume but a corresponding increase in vehicle speed. The town featured in the County Council's Town Centre Enhancement Programme aimed at the improvement of the environment and of trading viability.

Need for Measures: Speeds above the 30 mph limit and prevention of road accidents.

Measures Installed: A junction priority change at the south end of High Street at the Baldock Road junction and the provision of a mini roundabout at the north end of High Street, at the Vicarage Road junction. A variety of traffic calming features in between.

Special Features: Removal of all BT overhead cables. All measures designed to match Conservation Area status. Four 3 m x 8 m long throttle humps in High Street. Speed tables at Church Street and Norfolk Road. Combined carriageway/footway in Church Street defended by street furniture. No road markings except on humps. Disabled bay on footway.

Consultation: Four public exhibitions of four days each. Continuous consultation via Action Group of local interested parties for 3.5 years at intervals of 4-12 weeks according to needs.

Monitoring	Accidents	Speeds (mph)	Flows (16hr)
Before (after bypass)	7 in 2.5 years	35	7,300
After	1 in 3 yrs	24	6,850

Hertfordshire COUNTY COUNCIL
Transportation

Contact: Peter Dodd   Tel: 01992 556091

Authority: Hertfordshire County Council

Technical Data:

Location Type: Village centre, shops.

Road Type and Speed Limit: On C170, formerly A10 London-Cambridge trunk road.

Scheme Type: Town Centre Enhancement with traffic calming, plateau humps combined with single way throttles, wide footways.

Length of Scheme in Total: 350 m.

Dimensions: Height: N/A.
Width: Main carriageway 4.8 m (5 tonne weight limit southbound).
Ramp gradient: 1:18.
Distance from junctions: First throttle hump 30 m (south), 40 m (north).

Materials: Granite kerbs: 50 mm face granite sett channel.

Carriageway: Brown tarmacadam.

Footways: Cairnstone concrete slab paving in High Street.
York stone in Church Street and High Street junction.
Clay and concrete pavers in coaching accessways.
Antique concrete pavers in formed parking bays

Street furniture: Co-ordinated street furniture. Steel bollards. Wood seats.

Signs: Parking zone: 'Park in Marked Bays Only'. No yellow lines.
South bound: '5 Tonne Weight Limit'.
No carriageway centre line.

Lighting: New lighting scheme off facade of buildings.

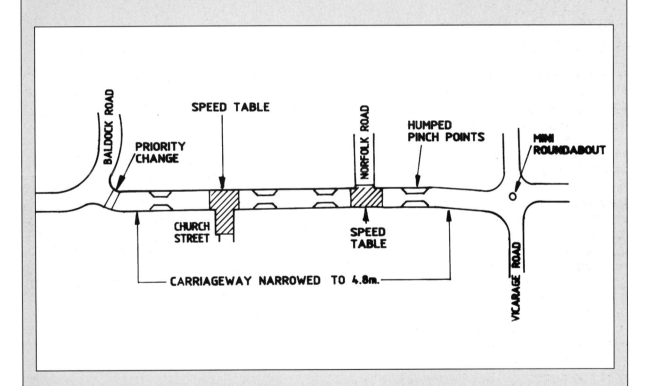

Cost: £400,000.

Contact's Comments: The speed of vehicles through the High Street was reduced and footways widened considerably to maximise the benefit from the construction of the bypass.

23 GATEWAYS, PINCH POINTS
Hertfordshire: Old Knebworth

Location: Hertfordshire: Old Knebworth village.

Implemented: January 1992.

Background: Experimental traffic calming scheme implemented within a small village of approximately 100 properties. Road through village often used as a peak hour commuter through route.

Need for Measures: Concern by residents over safety, owing to speed of through traffic.

Measures Installed: Three 'pinch points' with castellated blockwork surface. Timber baulks erected vertically to provide visual gateway to village.

Special Features: Castellated blockwork at pinch points, to provide 'rumble' type effect.

Consultation: Exhibitions, leaflets and consultation at Parish Council, District and County Council level.

Monitoring	Accidents (pia)	Speeds (mph)	Traffic (16 hr)
Before	2 in 3 years	42	2,000
After	0 in 2 years	39	1,800

Hertfordshire COUNTY COUNCIL
Transportation

Contact: Paul Irons   Tel: 01992 556056
Authority: Hertfordshire County Council

Technical Data:

Location Type: Small village.

Road Type and Speed Limit: Rural 'C' road: 60 mph.

Scheme Type: Pinch points with castellated blockwork and gateway feature.

Length of Scheme in Total: 800 m.

Dimensions: Difference in height of blocks to create castellated effect: 7 mm.
Width: 5.5 m.
Length: 7.0 m.
Ramp gradient: N/A.
Distance from junctions: Varies.

Materials : Throttles: Concrete block paving.
Kerbs: Marshalls conservation kerb.

Signs: 'Uneven Road' sign: Diag. 556.
'Road narrows': Diag. 517.
Road markings: for cycle lane.

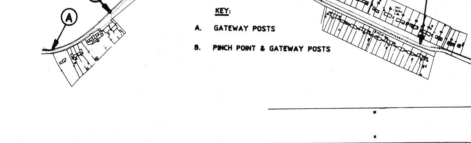

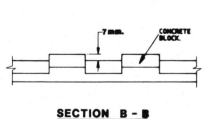

SECTION B - B

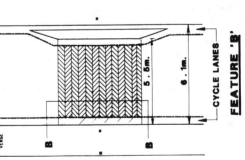

Lighting: Existing lighting columns upgraded with Victorian style lanterns.

Cost: Hertfordshire County Council £6,000.
North Hertfordshire District Council £6,000.
Private donation £6,000.
Total £18,000.

Contact's Comments: The experimental gateways have been of limited benefit, but will be built on using further pinch points within the village to enable a self enforcing 30 mph restriction to be imposed.

24-25 EXPERIMENTAL & REVISED SCHEME
Hertfordshire: Stanstead Abbotts

Left: (2) Revised scheme

Below: (1) Experimental scheme

Location: Hertfordshire: Stanstead Abbotts, High Street.

Implemented: (1) Experimental scheme April 1991.
(2) Revised scheme January-August 1992.

Background: Stanstead Abbotts bypass was opened in October 1987. The existing road width of 7.3 m had encouraged high speeds. Parking also needed to be revised to suite new status of road and greater consideration needed to be given to pedestrians. The physical state of footways and carriageway was very poor. Improvement of the general environment was to be considered.

Need for Measures: To reduce traffic speeds.

Measures Installed: (1) Installation of mini roundabout. Installation of seven experimental humps with varying heights and gradients, all to DOT standards. **(2)** Replaced by comprehensive enhancement of street environment by use of narrowed carriageway and reduced bell mouths. Blocked paving to carriageway. Central calming islands of 3 metre running width.

Special Features: (1) Road narrowing permitted installation of parking bays. **(2)** Central 1 metre wide strip in contrasting blocks to visually narrow carriageway. Raised features in contrasting blocks to assist pedestrian crossing movements and to disrupt visual continuity of carriageway.

Consultation: Very comprehensive. Exhibition held to determine public ideas and concerns. Exhibition held to show proposals and allow public to discuss scheme. Regular newsletters delivered to all homes. Regular Action Group meetings to discuss progress with public representatives.

Monitoring	Accidents (pia)	Speeds (85%ile mph)	Traffic (16 hr)
(1) Experimental scheme			
Before	6 in 3 years	31	7,100
After	Temp Installation	16	7,100
(2) Revised scheme			
Before	6 in 3 years	31	7,100
After	0 in 18 months	22	7,000

Hertfordshire COUNTY COUNCIL

Contact: Peter Dodd   Tel: 099255 6091
Authority: Hertfordshire County Council

Technical Data:

81

Location Type: Village residential and shopping. Conservation area.

Road Type and Speed Limit: Previously A414 primary route, downgraded to B181 secondary distributor: 30 mph.

(1) Scheme Type : Installation of mini roundabout. Temporary installation of 7 road humps of various heights all to DOT standards.

Length of Scheme in Total: 500 m.

Dimensions:	Height: 50mm, 75mm, 100mm
	Width: 7 m.
	Length: 6 m.
	Ramp gradients: 1:12, 1:18, 1:24.
	Distance from junctions: 30 m.
Materials:	Temporary road humps installed in bitumen macadam.
Signs:	'Road humps ahead' signing at extremities of scheme.
Lighting:	Layout of street lighting to enhancement scheme designed to take account of hump locations.

(2) Scheme Type: Narrowing of carriageway, reduction of bell mouths. Widening of footways, addition of parking bays. Block paving to carriageway, central contrast strip. Raised features for pedestrian crossing places.

Length of Scheme in Total: 360 m.

Dimensions:	No DOT Road Humps included.
Materials:	Concrete and clay block paving to carriage way and parking bays. Conservation kerbs laid flush to emphasise central contrast strip. Cast iron street furniture.
Signs:	No special signing.
Lighting:	Street lighting redesigned to suit new status of road. Standard columns enhanced with clamp on kit.

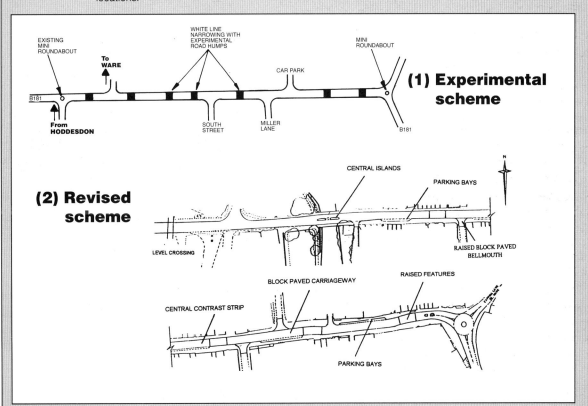

(1) Experimental scheme

(2) Revised scheme

Cost:	**(1)** Permanent mini roundabout £70,000. Temporary road humps £12,000. Total £82,000.	
	(2) £476,000.	

Contact's Comments: (1) The experimental scheme was carried out to calm traffic in village following opening of bypass. Very successful in reducing speed, but removed due to bad vibration effects on adjacent old properties. **(2)** The permanent scheme which replaced it has produced very successful speed reduction without the use of road humps in the conventional sense.

26 ENVIRONMENTAL ENHANCEMENTS
Kent: Sarre

Location: Kent: Sarre village.

Implemented: June 1993.

Background: Sarre village is a small community situated at the junction of the A28 and A253 roads on the Isle of Thanet. Prior to the works it was subject to a 40 mph limit. Existing street lighting was to full highway standards. Sarre has a population of approximately 100, many of whom are elderly. The central core of the village includes two pubs and a shop.

Need for Measures: Speeding, which led to accidents, and to a community split by heavy traffic flows.

Measures Installed: 30 mph speed limit. Entry gateways, refuge islands, widened footways, a chicane and central mini-roundabout. Closure of a minor link road. General environmental enhancement, including tree planting.

Special Features: Main road scheme based upon Danish style traffic calming for villages. Consideration given to scheme blending in with historic setting.

Consultation: With the District and Parish Councils, residents, police, bus companies, emergency services and local companies.

Monitoring	Accidents (pia)	Speeds (mph)	Traffic Flow
Before	11 in 3 years	47	up to 10,000
After	0 in 8 months	34	up to 10,000

Contact: Theresa Trussell Tel: 01622 696876
Authority: Kent County Council

Kent County Council
HIGHWAYS & TRANSPORTATION

Technical Data:

Location Type: Small village.

Road Type and Speed Limit: Two rural 'A' class roads, which previously joined at a priority junction in the central core. 40 mph speed limit reduced to 30 mph.

Length of Scheme in Total: On the A28: 510 m. On the A253: 350 m.

Dimensions: Gateways: Lane widths reduced to 3 m.
Islands: Lane widths reduced to 3 m.
Chicane: Central island and clipped kerb.
Mini roundabout: Sited at central junction.
Footways: Widened to at least 2 m.

Materials: Central core in Redlands Red Mac.
Footways and gateways in Halford & Baggeridge's brick paviours.
Bollards: Townscape timber.

Signs: Ringways: '30 mph' backed by yellow.

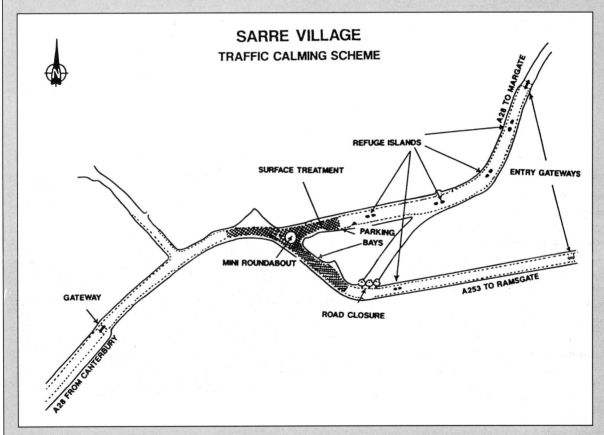

SARRE VILLAGE
TRAFFIC CALMING SCHEME

Cost: £140,000.

Contact's Comments: Kent's first 'A' road traffic calming scheme.

27 NARROWING, REFUGES
Kent: Gravesend

Location: Kent: Gravesend, A226 Shorne Crossroads to Thong Lane.

Implemented: September 1991.

Background: This section of the A226 is a secondary route between Strood and Gravesend. It is four lanes wide, subject to a 60 mph speed limit and is some 2.5 km in length.

Need for Measures: High speeds leading to numerous accidents, many severe. Seven fatal accidents occurred in the ten years between 1980 and 1990.

Measures Installed: Reduction of the number of lanes, by installing central refuge islands and remarking the road with one traffic lane and one cycle lane in each direction.

Special Features: Vibraline used to separate the cycle and traffic lanes.

Consultation: With the District Council and the police.

Monitoring	Accidents (pia)	Speeds	(mph)	Traffic (12 hr)
		west	east	
Before	15 in 3 years	61	59	9,400
After	6 in 30 months	55	50	9,400

Contact: Colin Martin   Tel: 01622 696823
Authority: Kent County Council

Kent County Council
HIGHWAYS & TRANSPORTATION

Technical Data:

Location Type: Rural four lane road.

Road Type and Speed Limit: Rural 'A' class road, secondary route: 60 mph.

Length of Scheme in Total: 2.5 km.

Dimensions: East bound widths: Cycle lane 1.5 m.

Traffic lane east bound 3.5 m.

Width of hatched area with traffic islands 2.0 m.

West bound widths: Traffic lane 3.5 m.
Cycle lane west bound 1.5 m.

Signs: Vibraline marking, 'keep left' arrows.

Lighting: Refuge beacons mounted on islands.

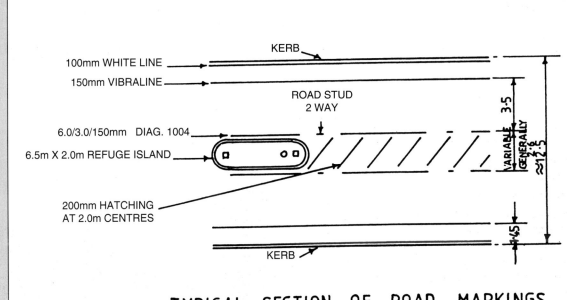

KERB

100mm WHITE LINE

150mm VIBRALINE

ROAD STUD
2 WAY

6.0/3.0/150mm   DIAG. 1004

6.5m X 2.0m REFUGE ISLAND

200mm HATCHING
AT 2.0m CENTRES

VARIABLE 3·5
GENERALLY 3·6 ≥12·5

1·45

KERB

TYPICAL SECTION OF ROAD MARKINGS
SCALE 1:250

Cost: £75,000.

Contact's Comments: None supplied.

28 ROAD HUMPS, REFUGES, MINIS
Kent: Hoo St Werburgh

Location: Kent: Hoo St Werburgh.

Implemented: November 1992.

Background: Main Road, Hoo St Werburgh is the principal road through the village. It is some 1,500 metres in length, subject to a 30 mph speed limit in part, complete with street lighting. It is bounded by schools, housing, a swimming pool and a recreation ground.

Need for Measures: Excessive speeding past schools, swimming pool, recreation ground, etc, which are constantly used by pedestrians, especially school children.

Measures Installed: 9 two-way flat humps, 4 mini roundabouts, 3 pedestrian refuge islands and 3 traffic islands. The school entrances are being converted to a one-way system and general kerb re-alignment has taken place. Weight restrictions to 7.5 tonne have been imposed with all the appropriate traffic signs. Existing lighting has been enhanced by 8 metre and 6 metre columns.

Special Features: Soft landscaping of wide grass verges.

Consultation: With police, Parish Council, local schools, Ambulance Service, Fire Brigade, bus operator and a period of public consultation.

Monitoring	Accidents (pia)	Speeds (mph)		Traffic	
		north	south	north	south
Before	15 in 3 years	44	45	4,400	4,100
After	0 in 18 months	29	24.5	4,920	4,850

Contact: Theresa Trussell Tel: 01622 696876
Authority: Kent County Council

Kent County Council
HIGHWAYS & TRANSPORTATION

Technical Data:

Location Type: Village local distributor road.

Road Type and Speed Limit: Urban classified 'C' single carriageway: 30 mph.

Length of Scheme in Total: 1600 m.

Dimensions: Height: 100 mm.
Width: 5 m.
Length: Plateau 10 m. Ramp 1 m.
Ramp gradient: 1:10.
Distance from junctions: Varies.

Materials: Plateau: Red DBM with 2 m.central band of black.DBM to denote crossing point.
Ramps: Concrete block paving coloured.buff laid in herringbone pattern on 50 mm sand bed.
Kerbs: Pre-cast concrete bull-nosed kerbs.
Street Furniture: Steel bollards.

Signs : Road Hump Signs: Diags. 516 and 557.2. Roundabout: Diags. 611.1 and 610.
Warning Signs: Diags. 557.1, 557.2,.516,.602

Lighting : Existing lighting upgraded to 6 m and 8 m columns in accordance with BS 5489.

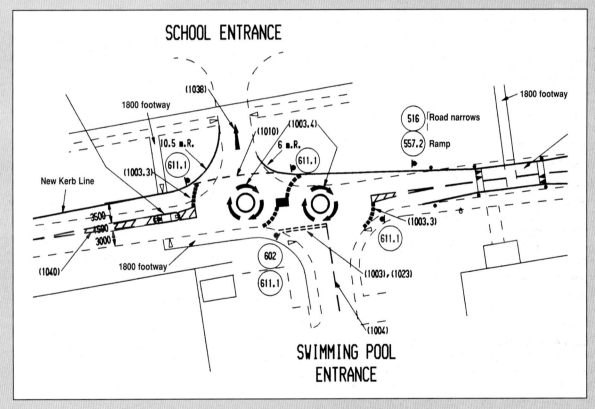

Cost: Kent County Council £70,000.
City of Rochester Upon Medway £230,000.
Total £300,000

Contact's Comments: None supplied.

29 ENVIRONMENTAL, GATEWAYS, CHICANES
Kent: Brasted

Location: Kent: Brasted village.

Implemented: November 1993.

Background: The village of Brasted lies on the A25 in West Kent, approximately 4 km from Kent's county boundary with Surrey. It has an estimated population of 1,500. Although effectively by-passed by the M25, traffic flows are heavy, up to 14,000 vehicles per day. Brasted is a typical Kentish village situated in a broad street with fine buildings of a variety of periods. It is subject to a 30 mph speed limit, but was unlit prior to the work apart from some low level amenity lighting.

Need for Measures: Speeds, which led to accidents involving pedestrians, including a number of fatalities.

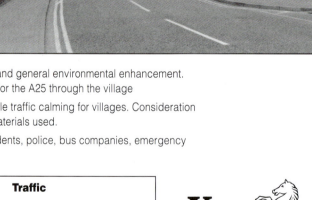

Measures Installed: Introduction of highway standard lighting; 'gateway' treatment; chicanes; change in carriageway colour through the central core; widened footways and general environmental enhancement. This work was incorporated into a major reconditioning scheme for the A25 through the village

Special Features: Main road scheme based upon Danish-style traffic calming for villages. Consideration given to scheme details particularly style of lighting and other materials used.

Consultation: With the Parish and District Councils, local residents, police, bus companies, emergency services and local businesses.

Monitoring	Accidents (pia)	Speeds (mph)	Traffic
Before	9 in 3 years	40	13,500
After	1 in 5 months	35	13,500

Contact: Theresa Trussell   Tel: 01622 696876
Authority: Kent County Council

Kent County Council
HIGHWAYS & TRANSPORTATION

Technical Data:

Location Type: Village.

Road Type and Speed Limit: 'A' road: 30 mph.

Length of Scheme in Total: 800 m.

Specification: Gateways: Sited at speed limit terminals,
using change in carriageway surface, marked posts, yellow backed signs.

Chicane: Sited at eastern gateway and in west part of village.

Carriageway treatment: Central core of village treated with 'red mac'.

Footways: Block paved and widened - parking bays installed.

Signs: Backed with yellow.

Lighting: Brought up to highway standard.

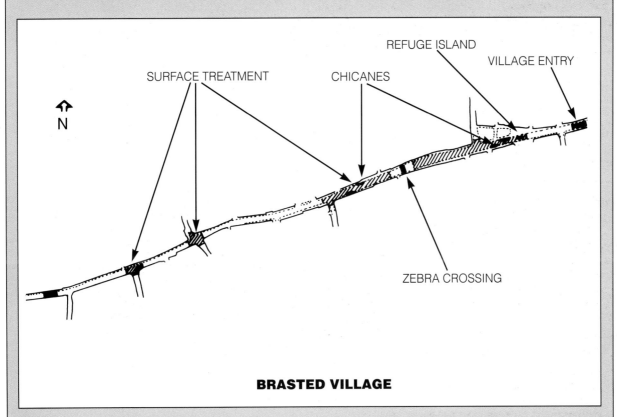

BRASTED VILLAGE

Cost: £425,000.

Contact's Comments: Residents had an overly optimistic view of the speed reductions to be expected, which led to some initial disappointment and problems. Scheme has now settled in.

30 GATEWAYS, NARROWING
Leicestershire: A427 Five Villages

Location: Leicestershire: A427, Walcote to Lubenham.

Implemented: March 1993.

Background: The A427 is the principal route that links Lutterworth and Market Harborough, passing through the villages of Walcote, North Kilworth, Husbands Bosworth, Theddingworth and Lubenham. Prior to the opening of the A14 in July 1994 it formed part of the primary route between the West Midlands and the East Coast.

Need for Measures: To slow traffic for speed limit compliance and to reduce accidents.

Measures Installed: Village gateway treatment with road narrowing.

Special Features: Trief kerbing with hatching to reduce carriageway widths.

Consultation: Parish Councils and emergency services.

Monitoring	Accidents (pia)	Speeds (85%ile mph)	Traffic (12 hr av)
Before	32 in 3 years	50.5	10,197
After	10 in 15 months	44.5	–

LEICESTERSHIRE COUNTY COUNCIL

Contact: Mike Hay   Tel: 0116 265 7220
Authority: Leicestershire County Council

Technical Data:

Location Type: Village approaches.

Road Type and Speed LImit: Rural classified: 30 mph.

Scheme Type: Gateways to villages with associated narrowings.

Length of Scheme in Total: At entrances to each village.

Dimensions: Width of carriageway: 6.50 m.
Width of hatching: 1.00 m.
Taper of kerbline (entrance): 1:30.
Taper of kerbline (exit): 1:10.

Materials: Walls: Engineering reds with mass concrete.

Kerbs: Trief type.

Signs: Illuminated with yellow background, Class 1 finish.

Signs: Diags. 744.1, 575, 569.1, 1013.1 and 1040. 1969 Regs Diags 1 and 2.

Lighting Extended in some cases to ensure good illumination of the narrowing.

Cost: £20,000 from Traffic Calming budget. £30,000 from Local Safety Schemes budget. Total £50,000.

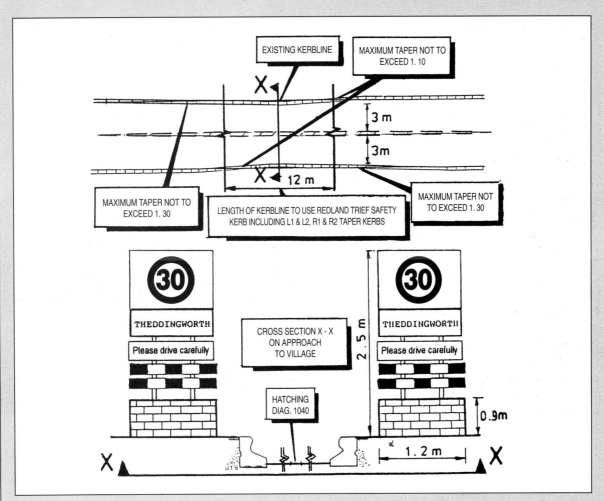

Contact's Comments: Initial monitoring indicated a general speed reduction of some 12%.
The newly opened A14 will dramatically affect the nature of the A427. However, at present it would appear that the 85 percentile speeds have remained stable.

31 ROAD HUMPS, MINI ROUNDABOUTS
Oxfordshire: Kennington

Location: Oxfordshire: Kennington village.

Implemented: May 1991.

Background: Kennington is a sizeable village lying just south of the Oxford Southern Bypass. Kennington Road/The Avenue forms the main road in the village. It provides a fairly direct alternative route to the A34(T) for traffic between Abingdon and Oxford, and was used by significant rat running traffic. Although this resulted in congestion in the morning peak, surveys of traffic speeds indicated a generally high level of abuse of the 30 mph speed limit.

Need for Measures: Poor accident and casualty record.

Measures Installed: Road hump system comprising 13 round top humps, 3 flat top humped zebra crossings and 3 flat top humps with mini-roundabouts.

Special Features: Mini-roundabouts were chosen to form the entry to the road hump system, but due to budget limitations it was not possible to achieve designs for these junctions that were judged to provide fully adequate deflection. To help overcome
this problem, it was decided to construct the roundabouts on extended flat top humps.

Consultation: With police, emergency services, bus operator, District Council, Parish Council and public (through public display)

Monitoring	Accidents (pia)	Speeds (mph)	Traffic Flow (16 hr)
Before	12 in 3 years	37	8,400
After	5 in 3 years	28	6,300

OXFORDSHIRE
COUNTY COUNCIL
Road Safety
CARING COUNTYWIDE

Contact: Ann Mortlock Tel: 01865 810400
Authority: Oxfordshire County Council

RURAL

Technical Data:

Location Type: Village.

Road Type and Speed Limit: Classified unnumbered: 30 mph.

Scheme Type: Road humps incorporating round top humps. Flat top humps with zebra crossing.
Flat top humps with mini roundabouts.

Length of Scheme in Total: 1.5 km.

Dimensions: Height: 90 mm (av).
Width: 6 m.
Length: Roundtop humps 3.7 m. Zebra crossing 4.6 m. Mini roundabout 15 m.
Ramp gradient 1:8.
Distance from junctions: Varies, up to 60 m.

Materials: Hot rolled asphalt.

Signs: Road hump warning signs (including humped zebra).
Roundabout/mini roundabout signing.
Temporary advisory signing.

Lighting: Additional lighting provided at mini roundabouts.

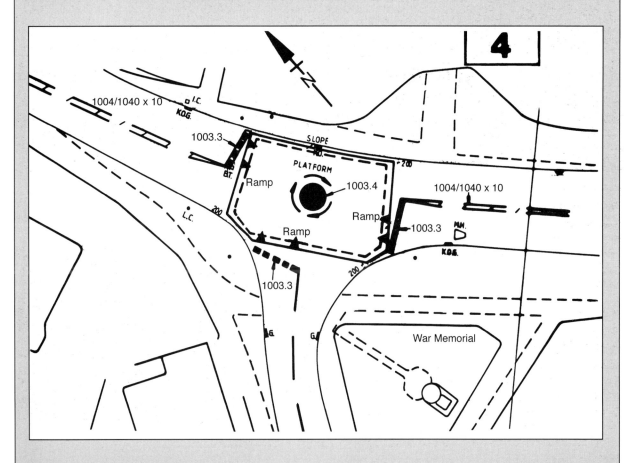

Cost: £55,000.

Contact's Comments: The measures have proved effective in reducing speeds and accidents. A large scale public attitude survey carried out approximately one year after completion showed a majority in support of the scheme. However, experience of later schemes suggests that the use of lower humps might have produced similar benefits without some of the adverse effects, particularly in relation to bus operators and emergency services.

32 GATEWAYS, CHICANES, RUMBLE STRIPS
Strathclyde: Croy

Location: Strathclyde: Croy, Constarry Road.

Implemented: April 1992.

Background: Croy is a former mining village to the north of Cumbernauld. Constarry Road is a classified road, B802, which provides a link between Cumbernauld and Kilsyth.

Need for Measures: To reduce pedestrian accidents on the road.

Measures Installed: Chicanes, gateways, pedestrian refuge islands,.thermoplastic rumble strips, clarification of lining and signing, pedestrian barrier rail and reflective concrete bollards.

Special Features: The use of thermoplastic rumble strips in conjunction with traffic calming elements.

Consultation: Emergency services, local, regional and district councillors, Member of Parliament and the Strathclyde Passenger Transport Executive.

Monitoring	Accidents (pia)	Speeds (mph)		Traffic (peak 2 way)
		Site A	Site B	
Before	11 in 3 years	37.5	49.0	5 - 600
After	1 in 1 year	33.5	40.5	5 - 600

Contact: Ken Aitken   Tel: 0141 227 2573

Authority: Strathclyde Regional Council

Strathclyde

Technical Data:

Location Type: Village.

Road Type and Speed Limit: Urban classified: 30 mph.

Scheme Type: Narrowings. Chicanes - single way working. Rumble strips.

Length of Scheme in Total: 1. 35 km.

Dimensions:
Chicane: Length 25 m. Reduced carriageway width 5.5 m.
Footway extensions 6 m radius extending to centre of carriageway.

Gateway: Length 13 m. Reduced carriageway width 4.5 m.
Footway extension width 1.4 m, including 0.5 m channel.

Pedestrian refuge island: Length 2 x 2 m islands with 2.5 m gap, width 1.3 m,
Reduced carriageway width: 3 m (both sides)

Thermoplastic rumble strip: Strips laid across carriageway with 500 mm channel for drainage
on both sides. Width 7.3 m in groups of 5 bars laid in a series on approaches to physical alterations,
with spacing between strips being 5 m and 10 m. Spacing is constant within a series.
Height 18 mm, width 100 mm, spacing 250 mm.

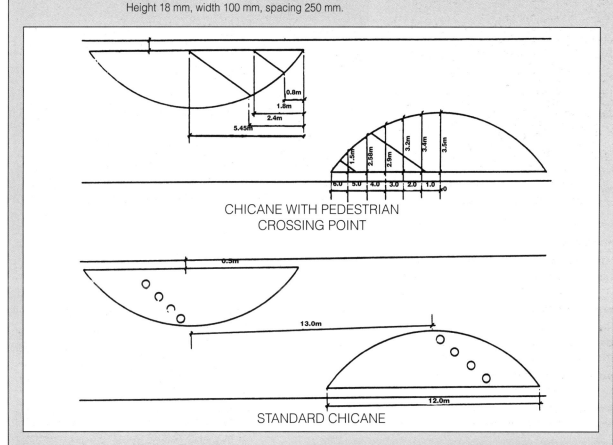

CHICANE WITH PEDESTRIAN
CROSSING POINT

STANDARD CHICANE

Materials:
Chicanes, gateways, islands: Asphalt with concrete kerbs and bollards.
Rumble strips: White thermoplastic material.

Signs:
Chicanes: Diags. 615 and 811.
Gateways: Diag. 516 with plate worded 'For 1 mile'.
Advisory 'Traffic calming scheme' sign - non prescribed.
Road markings: White lining at chicanes and gateways.

Lighting: As existing.

Cost: Strathclyde Roads Department Capital budget £15,000.

Contact's Comments: Although vehicle speeds at Site B are less than 'before' values, the particular form of rumble strip used at certain locations is not sufficiently effective as a speed reducing measure.

33 JIGGLE BARS
West Yorkshire: West Bretton

Location: West Yorkshire: West Bretton village.

Implemented: July 1988

Background: The A637 Huddersfield Road is a straight, high speed single carriageway road which runs through a rural area before it enters the village of West Bretton. It is approximately 9 metres wide on the approaches to the village, and subject to the national unrestricted speed limit. In the village itself the road narrows and contains a number of bends. Drivers tend to maintain their high speeds on entering the village and a number of of speed related loss of control accidents have occurred.

Need For Measures: To reduce vehicle speeds and hence speed related accidents within the village.

Measures Installed: Thermoplastic 'jiggle bars' on the approaches to the village.

Special Features: 'Jiggle bars' laid in a series of lateral bands to increase drivers' awareness by audible and visual warnings.

Consultations: With all emergency services and bus operators.

Monitoring		Speeds (mph)	
	Before	**After 1 week**	**After 1 month**
South east side			
To SE	50	33	42
To NW	47	35	40
North West Side			
To SE	45	37	39
To NW	44	32	38

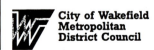

City of Wakefield Metropolitan District Council

Contact: Mr A M Salmon Tel: 01924 296066
Authority: Wakefield Metropolitan District Council

Technical Data:

Location Type: Village.

Road Type and Speed Limit: Rural 'A' class road: 60 mph outside village, 30 mph in village.

Scheme Type: Thermoplastic 'jiggle bars' and associated signing.

Length of Scheme in Total: Two sets of 49 'jiggle bars' each with a length of 45 metres.

Dimensions: Height: 8-12 mm.
                    Width: 8 m.
                    Length: 100 mm.

Materials: Orange thermoplastic material complying with BS 3262.

Signs: Signs to Diag. 562 and supplementary plate warning of road noise and vibration.

Lighting: Existing lighting in village retained.

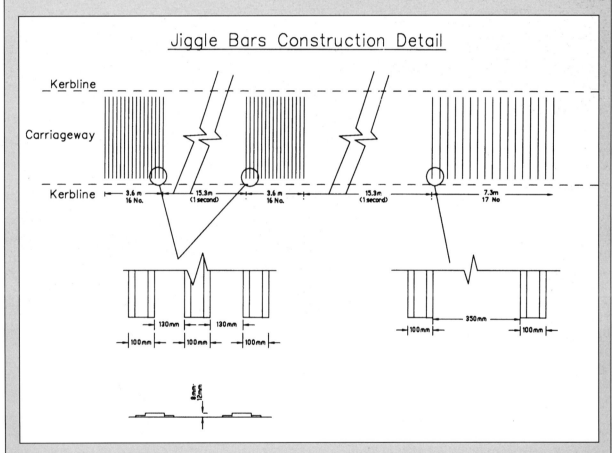

Jiggle Bars Construction Detail

Cost: £2,500.

Contact's Comments: Complaints were received from local residents regarding the noise the jiggle bars generated but it was felt that this was outweighed by the initial reduction in vehicle speeds. As time passed some motorists actually increased their speeds to minimise the effect of the bars whilst others negotiated them at extremely slow speeds. The resultant vehicle speed contrasts were considered hazardous and the bars were removed six weeks after implementation.

34 LANE NARROWING, CYCLE TRACK
West Sussex: Birdham

Location: West Sussex: Birdham village.

Implemented: March 1992.

Background: The A286 'Birdham Straight' is the only road into the Witterings, a busy coastal tourist and leisure area. The village of Birdham runs along the western side of this straight stretch of road which provides excellent forward visibility for motorists, thus inviting high speeds.

Need for Measures: Reduction of injury accidents resulting from high speeds and overtaking movements.

Measures Installed: A cycle track, central reserve markings and five central refuge islands.

Special Features: Southern footway replaced with a cycle track.

Consultation: Police, Parish Council.

Monitoring	Accidents (pia)	Speeds (85%ile mph)	Traffic (16 hr)
Before	13 in 3 years	56	11,350
After	10 in 326 months	44	–

west sussex county council

Contact: Stuart Smith   Tel: 01243 787921
Authority: West Sussex County Council

Technical Data:

Location Type: Village, coastal recreation area.

Road Type and Speed Limit: Rural classified: 60 mph (northern section) and 40 mph (southern section).

Scheme Type: A cycle track running along the eastern edge of the carriageway. A narrowing of the lane width by incorporating a lined central reserve and five refuge islands.

Length of Scheme in Total: 1.2 km.

Dimensions: Cycle track: Width 1.2 m.
                          Length: 1,200 m.

                    Carriageway: Width: 7.3-8.5 m.
                          Lane Width: 3.2 m (minimum).

Materials: Cycle track: Red surface treatment.

Signs: No additional signing.

Lighting: Illuminated bollards and beacons installed on the refuge islands.
                  Street lighting remained unchanged.

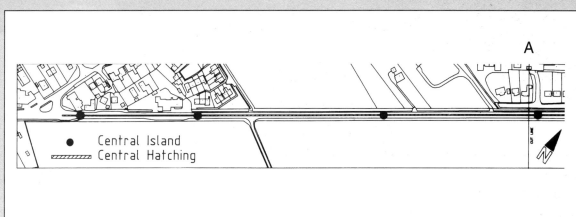

Cost: £50,000.

Contact's Comments: The construction of the cycle track accounted for a major part of the scheme cost.

35 NARROWING, FLAT TOP HUMPS, ENVIRONMENTAL TREATMENT
West Sussex: Bramber

Location: West Sussex: Bramber village.

Implemented: February 1993.

Background: On completion of the Bramber/Steyning bypass not all the through traffic transferred to the new route. An earlier traffic calming scheme using road narrowings proved unsuccessful in deterring through traffic and was removed. The measures subsequently implemented are outlined in this case study.

Need for Measures: Reduction of through traffic. Speed was not considered a problem.

Measures Installed: Six road humps, one carriageway width restriction and one chicane with paving to suit the environment.

Special Features: The scheme lies within a Conservation Area, which influenced its design and implementation. Reproduction period lamp columns and lanterns were installed. A block paving chicane with integral flower beds was incorporated into the village central area along with seating.

Consultation: Residents, Parish Council, District Council, police, emergency services, bus companies.

Monitoring	Accidents (pia)	Traffic (12 hr)	Through Traffic
Before	1 in 3 years	5,000	2,000
After	0 in 16 months	3,600	930

west sussex county council

Contact: John Dymond   Tel: 01906 74444
Authority: West Sussex County Council

Technical Data:

Location Type: Village centre.

Road Type and Speed Limit: Rural classified: 30 mph.

Scheme Type: Flat top road humps. A road narrowing incorporating a road hump.
A chicane formed from brick planters and bollards on a raised table in the village centre.

Length of Scheme in Total: 680 m.

Dimensions: Road humps: Height: 75 mm.
    Width: Full carriageway.
    Length: Plateau 2.5 m. Ramps 1 m.
    Ramp gradient: 1:13.

    Road narrowing/hump: Height: 75 mm.
    Width: 3.5 m.
    Length: Plateau 10 m. Ramps 1.2 m.
    Ramp gradient: 1:16.

Materials: Road humps: Rolled asphalt.

    Raised table/Chicane: Brindle/grey concrete block paving.
    Granite sett kerbing. Brick planters.Cast iron bollards.

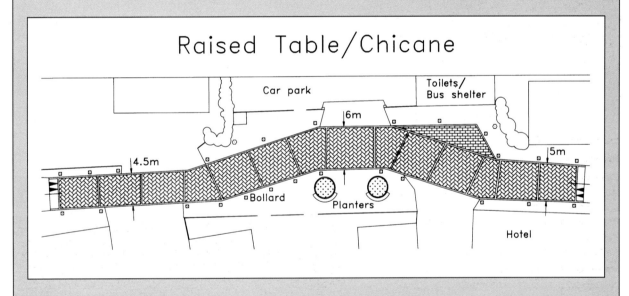

Raised Table/Chicane

Signs: Road hump Diag. 557.1. Road narrowing Diag. 517 signs.

Lighting: Reproduction period lamp columns and lanterns were installed
to complement the environmentally sensitive nature of the historic village setting.

Cost: £85,000. This includes £30,000 for special street lighting.

Contact's Comments: Successful. Almost half the through traffic was removed.

36 FLAT TOP HUMPS, MINI ROUNDABOUTS
West Sussex: Fontwell

Location: West Sussex: Fontwell.

Implemented: September 1992.

Background: Fontwell village straddled the A27 trunk road, Arundel Road, until a bypass around the village was built. A proportion of the through traffic remained, in the west bound direction, attracted by the shorter route and the speeds achievable due to the relatively light traffic flows.

Need for Measures: Reduce through traffic in the westbound direction and reduce the speed of the residual traffic.

Measures Installed: Four flat top humps and two mini roundabouts.

Special Features: The mini roundabouts were installed as speed reducing features at either end of the series of humps. The 40 mph speed limit was reduced to 30 mph to enable compliance with the road hump regulations.

Consultation: Police, emergency services, bus companies, local residents' organisation, District Council, local County Councillor.

Monitoring	Accidents (pia)	Speeds (mph)	Traffic (10 hr)	
			east bound	west bound
Before	1 in 3 years	40	825	475
After	0 in 21 months	25	605	480

west sussex county council

Contact: Tony Turley   Tel: 01243 777582

Authority: West Sussex County Council

Technical Data:

Location Type: Village centre.

Road Type and Speed Limit: Rural classified: 30 mph.

Scheme Type: One brick and three tarmac flat topped road humps, with bollards installed in the verge
adjacent to the toe of the ramps to draw attention to the presence of the humps.
Two mini roundabouts.

Length of Scheme in Total: 550 m.

Dimensions: Road humps: Height 75 mm.
    Width: Brick - kerb to kerb. Tarmac - tapered sides.
    Length: Plateau 2.5 m. Ramp 0.6 m.
    Ramp gradient: 1:8.
Mini roundabouts: Height: 100 mm at centre of dome.
    Diameter: 4 m.

Materials: Road humps: Three of rolled asphalt, one of clay bricks.
Bollards: Concrete.
Mini roundabouts: Rolled asphalt with reflective thermoplastic surface.

Signs: Road hump Diag. 557.1. Mini roundabout Diag. 611.1.
Replacement of the '40 mph' with '30 mph' speed limit signs.

Lighting: No change required to the existing street lighting.

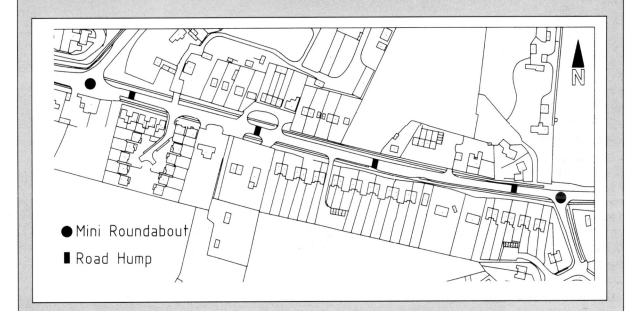

● Mini Roundabout

∎ Road Hump

Cost: £25,000.

Contact's Comments: The scheme proved successful in lowering speeds and reducing the
volume of west bound through traffic, particularly goods vehicles. The pelican crossing in the village
was removed as the flows no longer provided justification.

37-38 INITIAL AND REVISED SCHEME
West Sussex: Selsey

Location: West Sussex: Selsey village.

Implemented: (1) Initial scheme: February 1992.
(2) Revised scheme: March 1993.

Background: The B2145 Chichester Road is the only access to Selsey. The section of road into the village centre has houses along only one side where the 30 mph limit commences. The earlier set of traffic calming measures, based on brick gateways and rumble strips, failed to reduce speeds and gave rise the complaints from residents about excessive traffic noise. The revised scheme made use of road markings.

Need for Measures: Reduction of road accidents arising from excessive speeds.

Measures Installed: (1) Three gateways and four sets of mini road humps. **(2)** Carriageway narrowing by means of road markings.

Special Features: (1) Experimental use of mini road humps. **(2)** Coloured reflective road studs installed with the road markings.

Consultation: Police and Parish Council for both schemes, plus adjacent residents for revised scheme.

Monitoring	Accidents (pia)	Speeds (85%ile mph)	Traffic (16 hr)
(1) Initial scheme			
Before	4 in 3 years	42	10,100
After	0 in 1 year	39	10,100
(2) Revised scheme			
Before	4 in 3 years	40	10,100
After	1 in 16 months	32	10,100

west sussex county council

Contacts: Stuart Smith   Tel: 01243 787921

Authority: West Sussex County Council

Technical Data:

Location Type: Village approach

Road Type and Speed Limit: Rural classified: 30 mph.

(1) Scheme Type: Three block gateways with bollards located in the verge to highlight their presence. Four sets of mini humps/rumble strips, each made up of four round top mini road humps.

Length of Scheme in Total: 500 m.

Dimensions: Mini Road Humps: Height: 20-30 mm.
                    Width: 500 mm. Spacing: 1500 mm.

                    Gateways: Height: Existing carriageway
                    level. Width: Full carriageway. Length:
                    10 m.

Materials: Mini humps: Rolled asphalt.

                    Gateways: Block paving bordered by pre
                    cast concrete channel blocks.
                    Bollards: Glasdon 'Admiral'.

Signs: No additional signing.

Lighting: No change was made to the existing street
                    lighting.

(2) Scheme Type: Reducing the effective lane width by adding carriageway edge and central reserve markings as well as the demarcation of residents' parking spaces. Three brick gateways installed as part of an earlier traffic calming scheme were retained.

Length of Scheme in Total: 330 m.

Dimensions: Carriageway: Width: 8.6 m. Lane width:
                    3 m. Lay by width: 1.9 m.

                    Gateways: Height: Existing carriageway
                    level. Width: Full carriageway.
                    Length: 10 m.

Materials: Gateways: Block paving bordered by
                    pre-cast concrete channel blocks.

Signs: No additional signing.

Lighting: No changes were made to the existing
                    street lighting.

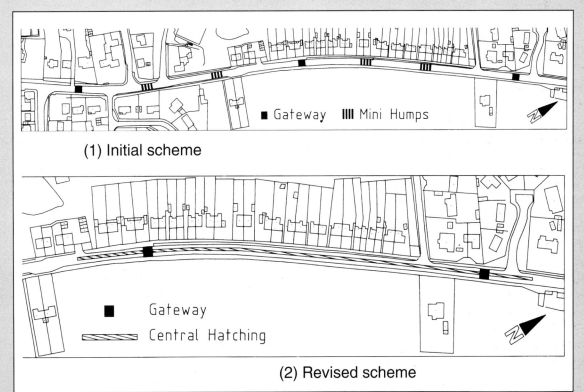

(1) Initial scheme

■ Gateway IIII Mini Humps

(2) Revised scheme

■ Gateway

▨▨▨ Central Hatching

Cost: **(1)** £ 9,500. **(2)** £ 3,000.

Contact's Comments: The earlier scheme failed because the reduction in speeds recorded initially was only temporary and the residents complained about excessive noise caused by vehicles crossing the humps. The mini humps were removed. The modified scheme has successfully reduced traffic speeds.

Residential

Case Studies 39 – 85

39 20 MPH ZONE, VARIED MEASURES
York, The Groves

Location: North Yorkshire: York, The Groves.

Implemented: Measures: April 1991.
20 mph zone: March 1992.

Background: The Groves is a large residential area on the edge of York's historic City Centre. Lowther and Penley's Grove Streets are both one-way and bus routes. Park Grove and Neville Terrace were commuter cut throughs. Within the area are a school, shops and a number of businesses.

Need for Measures: The closure of Deangate, the road beside York Minster, led to an increase in traffic in The Groves. Residents petitioned the Council to do something about the increase in traffic and its speed.

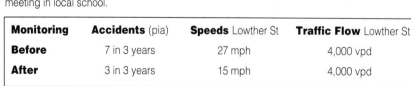

Measures Installed: 15 tapered edge road humps.
10 pavement build outs at gateways. 10 road humps with pavement build out. 6 raised junctions. 5 chicanes. 2 speed tables. 2 one way plugs. 1 road closure.
A resident only parking scheme was installed at the same time as the traffic calming.

Special Features: Block paving on all speed tables and raised junctions. 80 mm high measures on bus routes.

Consultation: Residents opinion survey before and after scheme. Meeting with resident associations. Letters to police, emergency services, bus company road users, disabled groups, local businesses and schools. Public meeting in local school.

Monitoring	Accidents (pia)	Speeds Lowther St	Traffic Flow Lowther St
Before	7 in 3 years	27 mph	4,000 vpd
After	3 in 3 years	15 mph	4,000 vpd

CITY OF YORK

Contact: Tim Pheby Tel: 01904 65347   **Graham Cressey Tel: 01609 780780**
Authority: York City Council   **North Yorkshire County Council**

Technical Data:

Location Type: Residential, adjacent to historic city centre.

Road Type and Speed Limit: Unclassified: 20 mph.

Type of Measure: Speed table and chicane.

Length of Scheme: N/A.

Dimensions:		Speed Table	Build Out
	Height:	80 mm	100 mm
	Width:	5 m	1.8 m max
	Length:	9 m	7 m
	Ramp gradient:	1:10	-

Materials: Plateaus: Grey blocks.
Ramps: White and charcoal blocks.
Kerbs: Standard bull nose concrete.
Bollards: Timber (appropriate to 'The Groves').

Signs: Diags. 674 and 675 at limits of 20 mph zone.

Lighting: Existing.

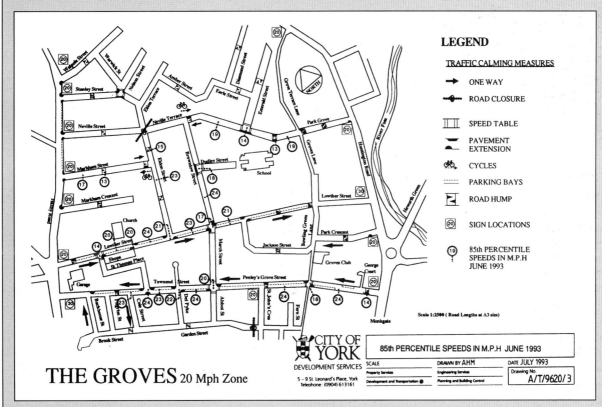

THE GROVES 20 Mph Zone

LEGEND

TRAFFIC CALMING MEASURES

→ ONE WAY

ROAD CLOSURE

|||| SPEED TABLE

PAVEMENT EXTENSION

CYCLES

PARKING BAYS

ROAD HUMP

SIGN LOCATIONS

85th PERCENTILE SPEEDS IN M.P.H JUNE 1993

Scale 1:2500 (Road Lengths at A3 size)

CITY OF YORK
DEVELOPMENT SERVICES
5 – 9 St. Leonard's Place, York
Telephone: (0904) 613161

85th PERCENTILE SPEEDS IN M.P.H JUNE 1993		
SCALE	DRAWN BY AHM	DATE JULY 1993
Property Services	Engineering Services	Drawing No.
Development and Transportation ●	Planning and Building Control	A/T/9620/3

Cost: £130,000.

Contact's Comments: None supplied.

40 20 MPH ZONE, CHICANES, NARROWING, FLAT TOP HUMPS
Bradford, Scotchman Road

Location: West Yorkshire: Bradford, Scotchman Road and Jesmond Avenue.

Implemented: May 1991. 20 mph zone status 1993.

Background: Jesmond Avenue and Scotchman Road are typical examples of inner city streets subject to a 30 mph speed limit with sub-standard street lighting. Jesmond Avenue is entirely residential, consisting of terraced housing. Scotchman Road forms a direct access to two first schools and one middle school. Both are used as commuter cut throughs.

Need for Measures: Rat-running and excessive speeds which led to child pedestrian accidents.

Measures Installed: A system of chicanes, road narrowings and platforms.

Special Features: Consideration was given to the scheme being sympathetic to the local environment and to maximising available on-street parking in Jesmond Avenue by constructing features at junctions.

Consultation: With police, emergency services, residents and local schools.

Monitoring	Accidents (pia)	Speeds (mph)		Traffic (peak hr)	
		Scotchman Rd	Jesmond Av	Scotchman Rd	Jesmond Av
Before	4 in 3 years	32	27	679	101
After	3 in 3 years	19	17	604	57

Transportation and Planning
Directorate of Community & Environment
City of Bradford Metropolitan Council

Contact: J M Wallis   Tel: 01274 757401

Authority: City of Bradford Metropolitan Council

Technical Data:

Location Type: Urban residential.

Road Type and Speed Limit Urban unclassified: 30 mph.

Scheme Type: Flat top humps, narrowings and chicanes in various combinations.

Length of Scheme in Total: Scotchman Road 450 m. Jesmond Avenue 300 m.

Dimensions: Height: 80 mm.
Width: Varies min 4 m, max 6.2 m.
Length: Varies min 5 m, max 26 m.
Ramp Gradient 1:12
Distance between measures: Average 60-80 m.

Materials: Marshalls Lambeth deterrent paving flags. Marshalls 'VTO' Buff Saxon paving slabs.
Marshalls Wimslow reflective bollard

Signs: 20 mph speed limit signs.

Road Markings: Centre line markings Diag. 1004. Hatched markings Diag. 1040.
'Give way' markings Diag. 1003.

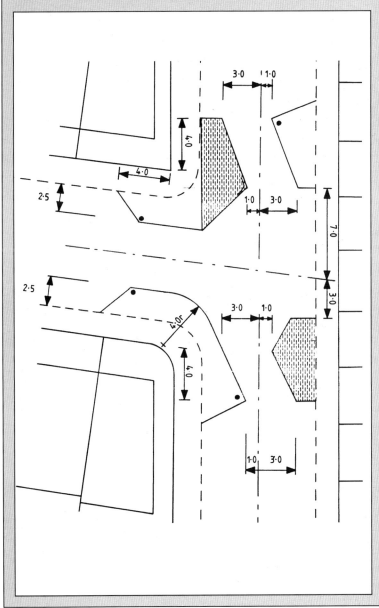

Lighting: New 8 m steel columns.

Cost: Department of the Environment Urban Programme Grant (100%) £105,000.

Contact's Comments: The three accidents since the scheme was installed all occurred within 6 months in 1994. Further investigation is required to determine any cause. For instance, are motorists becoming used to the measures?

41 20 MPH ZONE, ROAD HUMPS, RAISED JUNCTIONS
Ipswich, Britannia Area

Location: Suffolk: Ipswich, Britannia area.

Implemented: August 1992.

Background: The Britannia area is a residential area two miles to the east of the town centre. It was subject to a 30 mph speed limit and contains two schools. Britannia Road runs from north to south through the area and some rat-running took place along this road.

Need for Measures: The alignment of Britannia Road encouraged high vehicle speeds. There had been 21 injury accidents in 5 years, 19 of which involved the more vulnerable road users.

Measures Installed: 37 road humps and 3 speed tables. The whole area is subject to a 20 mph speed limit.

Special Features: The hump at Britannia Primary School is flat top and constructed in red blocks, as were the three speed tables.

Consultation: With local residents, local schools, emergency services and bus companies. A public exhibition was held at one of the schools.

Monitoring	Accidents (pia)	Speeds (mph)	Traffic Flow (1 hour)
Before	21 in 5 years	30	2,000
After	2 in 2 years	16	2,000

Contact: Ian Coleman   Tel: 01473 262825
Authority: Ipswich Borough Council and Suffolk County Council

IPSWICH
BOROUGH COUNCIL

Suffolk County Council

Technical Data:

Location Type: Residential area.

Road Type and Speed Limit: Urban unclassified: 30 mph.

Scheme Type: 36 round top humps and 1 flat top hump. 3 flat top speed tables.

Length of Scheme in Total: Area scheme: 1 km x 1 km approximately.

Dimensions: Height: 80 mm.
Width: 6 m
Length: Flat top 3.4 m. Ramps 1.2 m. Round top 3.7 m.
Ramp gradient: 1:15.
Distance from junctions: Not more than 40 m.

Materials: Round top humps: Red asphalt.
Flat Top Humps and speed tables: red block paving BDC Jessup ramps in yellow.

Signs: Road Hump: Diag. 557.
Special ' 20/30 mph' signs.
Special advanced map type '20 mph' signs.

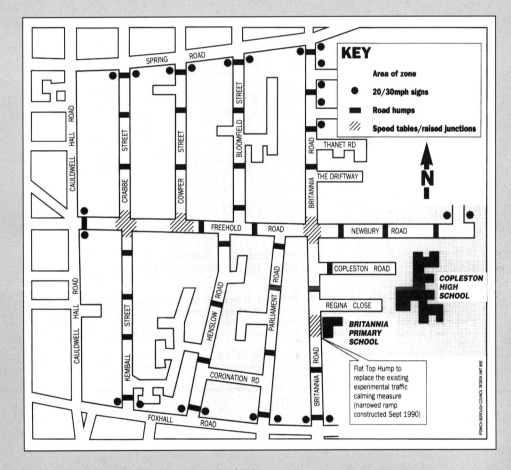

Lighting: Existing road lighting systems maintained.
The position of the street lights coincided sufficiently with the humps so that changes were not justified.

Cost: Suffolk County Council Transport Supplementary Grant £85,000.

Contact's Comments: The scheme appears to be working very well and no complaints have been received. The humps seem to have captivated the imagination of the Ipswich public as many similar requests for traffic calming have been received during the last two years. The scheme was made a permanent 20 mph zone in May 1994.

42 20 MPH ZONE, VARIED MEASURES
Manchester, Crumpsall Green

Location: Manchester: Crumpsall Green.

Implemented: Stage 1: April 1992. Stage 2: December 1993.

Background: Crumpsall Green is an area of north Manchester, bounded by three district distributor routes and consisting of Victorian terrace and pre-war semi-detached housing.

Need for Measures: The area was targeted due to the high level of injury accidents and the use of routes within the area as rat-runs.

Measures Installed: 20 mph speed limit order, road humps, speed tables, traffic signals, road narrowing, one-way working, waiting restrictions and road closures.

Special Features: Two stage implementation, the first involving the areas to the east and west of Lansdowne Road and Crescent Avenue, and the second introducing measures on these two streets.

Consultation: Residents, traders and local schools through public meetings and exhibitions, also liaison with emergency services.

Monitoring	Accidents (pia)	Speeds (mph[1])	Traffic (2 way 7 hr[2])
Before	44 in 5 years	N/A	2,500
After	N/A	10.2/14.0	1,116
1) Speeds - At hump/between humps , 2) Refers to flows on Lansdowne Road/Crescent Avenue			

Contact: Peter Widdall   Tel: 0161 247 3143
Authority: Manchester City Council

 **Manchester**
making it happen

Technical Data:

Location Type: Residential.

Road Type & Speed Limit: Urban 'C' and unclassified roads: 30 mph.

Scheme Type: 20 mph speed limit zone. Measures include flat and round top humps, road narrowing, one-way working, road closures and traffic signals.

Length of Scheme in Total: Approximately 0.5 sq km in area.

Dimensions: Height: 100 mm.
Width: 6.3 m.
Length: Plateau 5 m. Ramp 1 m.
Ramp gradient: 1:10.
Distance from junctions: 7 m.

Materials: Extended flat top humps: Red block paving.
Round top humps: Bituminous macadam.
Kerbs: Precast concrete.
Street Furniture: Recyled plastic bollards.

Signs: '20 mph Zone' and 'Zone Ends' signs: Diags. 674 and 675.
'Road Narrows' sign: Diag 517.
Road hump markings: Diags.1060 and 1061.
Road hatching marking: Diag. 1040.

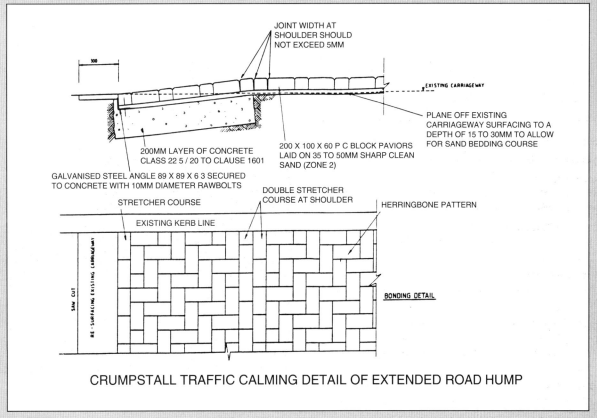

JOINT WIDTH AT SHOULDER SHOULD NOT EXCEED 5MM

EXISTING CARRIAGEWAY

PLANE OFF EXISTING CARRIAGEWAY SURFACING TO A DEPTH OF 15 TO 30MM TO ALLOW FOR SAND BEDDING COURSE

200MM LAYER OF CONCRETE CLASS 22 5 / 20 TO CLAUSE 1601

200 X 100 X 60 P C BLOCK PAVIORS LAID ON 35 TO 50MM SHARP CLEAN SAND (ZONE 2)

GALVANISED STEEL ANGLE 89 X 89 X 6 3 SECURED TO CONCRETE WITH 10MM DIAMETER RAWBOLTS

STRETCHER COURSE

DOUBLE STRETCHER COURSE AT SHOULDER

HERRINGBONE PATTERN

EXISTING KERB LINE

SAW CUT

RE-SURFACING EXISTING CARRIAGEWAY

BONDING DETAIL

CRUMPSTALL TRAFFIC CALMING DETAIL OF EXTENDED ROAD HUMP

Lighting: Upgraded where necessary, eg. for traffic signals.

Cost: Manchester City Council Local Safety Schemes allocation £175,000
(Eligible for Transport Supplementary Grant to cover 50% of the cost).

Contact's Comments: Prefer in future to consult via residents groups/organisations rather than hold a general public meeting which can attract all sorts of complaints and be counter productive.

43 20 MPH ZONE, VARIED MEASURES
Poole, Upper Parkstone

Location: Dorset: Poole, Upper Parkstone, Heatherlands.

Implemented: August 1993.

Background: The Upper Parkstone area has 1930s housing on narrow but straight residential roads. It is bounded by heavily trafficked main roads on all four sides which are subject to a 30 mph speed limit. The 1.7 sq km area is to be divided into three 20 mph zones, of which Heatherlands is the first to be completed with the others planned over the next few years.

Need for measures: The area has more injury accidents per km than any other residential area in Poole and suffered from rat-running by traffic avoiding busy junctions on the boundary main roads.

Measures Installed: 21 flat-topped 75 mm high plateaus with 1 in 12 approach slopes, 2 round top humps, 3 junction plateaus, 2 mini roundabouts, 12 protected parking areas and a 20 mph speed limit.

Special Features: The plateaus are of varied design including narrows, planters, granite setts at the top and bottom of the approach ramps, and the use of recycled plastic bollards.

Consultation: Extensive local public consultations took place including the emergency services and affected bus companies.

Monitoring	Accidents (pia)	Speeds (mph)	Traffic (AADT)
Before	25 in 3 years	27	5,333
After	0 in 9 months	16	-

DORSET County Council

BOROUGH OF POOLE

Contact: Tim Westwood   Tel: 01305 224938   Steve Dean   Tel: 01202 675151
Authority: Dorset County Council   Borough of Poole

Technical Data:

Location Type: Densely populated urban residential area with narrow streets laid out in straight lines but with steep gradients running north to south.

Road Type and Speed Limit: All unclassified: 30 mph.

Scheme Type: 20 mph zone with protected parking areas.

Length of Scheme: 4 roads comprising 2.2 km.

Dimensions: Plateaus: Height 75 mm. Width varies, some 3.5 m. Length 6 m. Ramp gradient 1:12. Spacing 50 m - 120 m.

Protected parking: Kerbed planters 2 m x 4 m.

Materials: Plateaus: Hot rolled asphalt with three rows of granite setts at each end. Ramps: Hot rolled asphalt with four rows of granite setts at the bottom. Bollards: Red recycled plastic.

Signs: 20 mph speed limit zone signs. No road markings, except for arrows on road hump approach slopes.

Lighting: Existing.

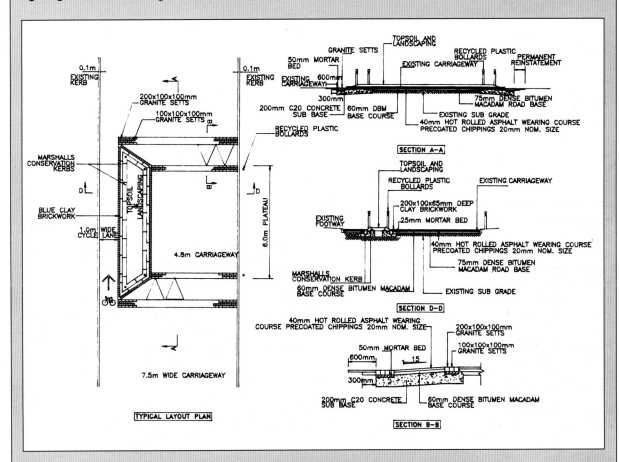

Cost: Dorset County Council - basic roadworks £224,500.
Dorset County Council - design and supervision £46,500.
Dorset County Council - consultation £7,000.
Poole Borough Council - environmental enhancements £41,500.
Total £319,500.

Contact's Comments: Approach ramps for flat top road humps changed from 1 in 15 to 1 in 12 as the former was considered too flat. The scheme was a finalist in the 1993 *Urban Street Environment* Traffic Calming Award competition.

44-45 20 MPH ZONE, ROAD NARROWINGS WITH SPEED CUSHIONS, CYCLE SLIPS
Nuneaton, Camp Hill Area

(2)

Location: Warwickshire: Nuneaton, Camp Hill area, **(1)** Cedar Road and **(2)** Queen Elizabeth Road.

Implemented: February 1994.

Background: Both Cedar Road and Queen Elizabeth Road are long roads serving the Camp Hill housing estate, a recently commissioned 20 mph zone. Cedar Rd has housing on both sides, with parking on-street and high pedestrian activity. It is 1 km long and relatively straight.. Queen Elizabeth Road is 2 km long, with a large green area on one side and housing the other. Both are frequent bus routes and main access routes for the emergency services. See also Case Study 46.

Need for Measures: To reduce the high accident record by slowing traffic speeds.

Measures Installed: (1) Series of four straight road narrowings with speed cushions and road humps. **(2)** Series of offset road narrowings with speed cushion and speed tables at junctions.

(1)

Special Features: Imprinted asphalt paving, speed cushion and cycle slip for both. Queen Elizabeth Road **(2)** has imprinted paving instead of block paving in channels, and uses the speed cushion off-set at an angle as a 'tracking' device.

Consultation: Full legal consultation, plus residents by leaflets and four public exhibitions.

Monitoring (1) Cedar Road	Accidents (pia)	Speeds (mph)	Traffic (16 hr)
Before	8 in 3 years	30 (85% ile)	3,000 (estimated)
After	0 in 2 months	20 (estimated)	3,000 (estimated)
(2) Queen Elizabeth Road			
Before	12 in 3 years	42 (85%ile)	2,500
After	0 in 1 month	27 (estimated)	2,500 (estimated)

Warwickshire
County Council

Contact: Paul Bellotti   Tel: 01926 412179

Authority: Warwickshire County Council

Technical Data:

(1) Location Type: Residential.

Road Type and Speed Limit: Urban unclassified: 20 mph.

Scheme Type: Straight road narrowing with speed cushion and cycle slips.

Length of Scheme in Total: 1 km.

Dimensions: Height: 65 mm. Width: 2.1 m.
Length: 4 m.
Ramp gradient: 1 in 10.
Distance from junctions: Varies.

Materials: Plateau and ramps: Red asphalt, imprinted asphalt shapes in channels.

Kerbs: Concrete.

Street Furniture: Bollards with retro-reflective strips.

Signs: None.

(2) Location Type: Residential.

Road Type and Speed Limit: Urban unclassified: 30 mph. Boundary road to a 20 mph zone.

Scheme Type: Offset road narrowing with speed cushion - single-way working.

Length of Scheme in Total: 2.2 km.

Dimensions: Height: 55 mm high speed cushion.
Width: 1 m. Length: 6 m.
Ramp gradient: 1 in 10.
Distance from junctions: Varies.

Materials: Plateau and Ramps: Red asphalt.

Kerbs: Concrete.

Street furniture: Bollards with retro-reflective red/white stripes.

Signs: None.

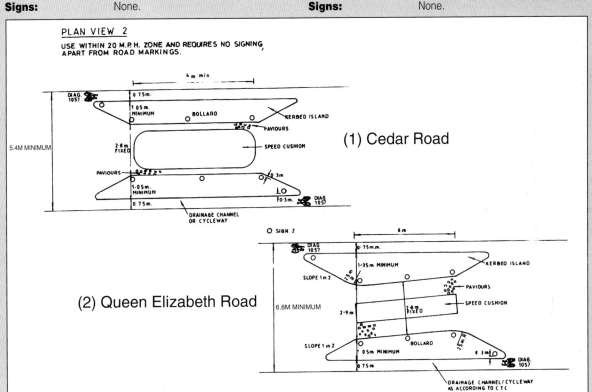

PLAN VIEW 2

USE WITHIN 20 M.P.H. ZONE AND REQUIRES NO SIGNING, APART FROM ROAD MARKINGS.

(1) Cedar Road

(2) Queen Elizabeth Road

Lighting: HP sodium lighting

Cost: £2,000 per straight road narrowing funded from Local Safety Scheme budget.

Lighting: LP sodium lighting

Cost: £2,000 per offset road narrowing, funded from Local Safety Scheme budget.

Contact's Comments: The single way working road narrowing slows drivers' speeds. **(1)** The circular horizontal profile of the cushions in Cedar Road means that wider vehicles go over a shallower hump. The minimum road width required ior this is 6.4 metres to allow parking both sides of the road, either side of the narrowing. **(2)** In Queen Elizabeth Road the narrow speed cushion forces drivers to follow the alignment of the narrowing and not take a straight (racing) line through the middle. This layout was trialed off road with all the emergency services and they prefer it to a road hump. Requires minimum road width of 6.8 metres. Camp Hill 20 mph zone was a finalist for the 1994 *Urban Street Environment* Traffic Calming Awards.

46 20 MPH ZONE, CHICANES, CYCLE SLIPS
Nuneaton, Whittleford Road

Location: Warwickshire: Nuneaton, Camp Hill area, Whittleford Road.

Implemented: April 1994.

Background: Whittleford Road is a local distributor road about 2 km in length, serving two large residential estates which form a newly commissioned 20 mph zone (see Case Studies 44-45). It is a bus route and the emergency services use it as an access road to these estates. It has an open road environment and speeding was a problem.

Need for Measures: To reduce high accident record by slowing traffic speeds to 30 mph.

Measures Installed: Two chicanes and speed tables at junctions.

Special Features: Cycle slips.

Consultation: Full legal consultation and with residents using leaflet drop with four public exhibitions.

Monitoring	Accidents (pia)	Speeds (85%ile mph)	Traffic (16 hr)
Before	33 in 3 years	41	9,000
After	1 in 2 months	30	8,000

Warwickshire County Council

Contact: Phil Cook   Tel: 01926 412345

Authority: Warwickshire County Council

Technical Data:

Location Type: Residential.

Road Type and Speed Limit: Urban unclassified: 30 mph.

Scheme Type: Full chicanes and speed tables at junctions.

Length of Scheme in Total: 2 km.

Dimensions:
Height: N/A.
Width: N/A.
Length: 20 m.
Ramp gradient: N/A.
Distance from junctions: 100 m.

Materials:
Plateau and ramps: N/A.
Kerbs: Concrete.
Street furniture: Bollards with retro-reflective red/white strips.

Signs: Yellow backed, lit 'Give Way' - class 1.

Lighting: Low pressure sodium lighting (as existing).

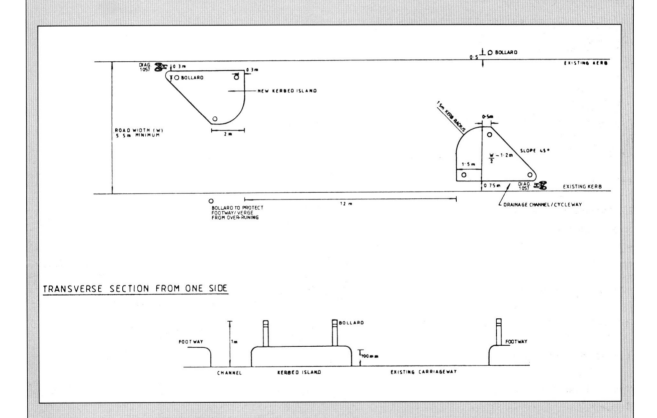

TRANSVERSE SECTION FROM ONE SIDE

Cost: £2,000 per chicane, funded from Local Safety Scheme budget. Total £4,000.

Contact's Comments: The single-way priority slows drivers, but at peak times queues can be in the order of 25 cars long. The chicanes were used as a substitute to road humps to benefit the emergency services. Residents prefer road humps because they do not have to give way to oncoming traffic, incurring delay. The chicanes have been removed due to complaints about congestion.

47 20 MPH ZONE, FLAT TOP HUMPS
Stoke-on-Trent, Fenton Area

Location: Staffordshire: Stoke-on-Trent, Beville Street and Vivian Road.

Implemented: November 1991 - April 1993.

Background: Fenton Community Renewal Area is a predominantly residential area, approximately 400 metres square. The major concern of the residents was the amount of traffic rat-running through the area to avoid long queues and delays experienced on the adjoining A50.

Need for Measures: Speed and amount of traffic rat-running through the area which led to child pedestrian and cyclist accidents.

Measures Installed: A series of 33 flat top road humps, followed by the introduction of a 20mph zone.

Special Features: In accordance with the area's Community Renewal status, its appearance was upgraded by specifying block paved flat top humps.

Consultation: With local residents, all emergency services. No bus route through the area.

Monitoring	Accidents (pia)	Speeds (mph)	Traffic Flow (peak)
Before	9 in 6 years	40	350
After	0 in 2.5 years	15	217

Staffordshire
County Council

Contact: Will Loomes   Tel: 01782 404146
Authority: Stoke-on-Trent City Council

Technical Data:

Location Type: Residential.

Road Type and Speed Limit: Urban unclassified: 20 mph.

Scheme Type: 20 mph zone. Flat top humps.

Length of Scheme in Total: 2.4 km.

Dimensions: Height: 100 mm.
Width: Varies.
Length: Plateau 2.5 m. Ramp 600 mm (lengthened to 1.0m).
Ramp gradient: 1:6 (reduced to 1:10)

Materials: Plateau: Red concrete block paving.
Ramps: Stone setts, tarmac.
Kerbs: Concrete.
Bollards: Plastic - cast iron lookalikes.

Signs: '20 mph zone'.

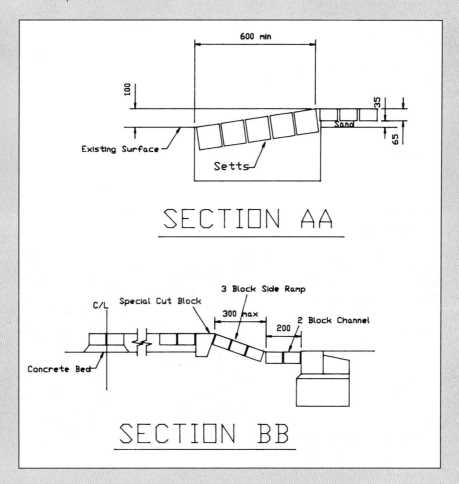

Lighting: Steel columns.

Cost: City of Stoke-on-Trent Environmental Improvements budget £70,000.
Staffordshire County Council Traffic Calming budget £8,000.
Total £78,000.

Contact's Comments: The scheme has been very effective in reducing speeds and accidents. Well received by residents. Shortly after implementation, certain stretched funeral limousines were grounding on the humps approaching Fenton Cemetery. The profile of the humps on this particular route had to be amended to resolve this problem.

48 FLAT TOP HUMPS, JUNCTION PLATEAU
Bristol, Throgmorton Road

Location: Avon: Bristol, Throgmorton Road.

Implemented: November 1993.

Background: Throgmorton Road is a residential street 620 metres in length, which provides access to a junior and infants school.

Need for Measures: To slow traffic speeds, deter rat running in a residential street and to reduce pedestrian accidents.

Measures Installed: A series of eight plateau topped humps (including a humped zebra crossing), and a raised junction plateau.

Special Features: 22 buses an hour, therefore used 6 metre plateau with 1 metre ramp to minimise the amount of discomfort felt by bus passengers, as well as to achieve a significant reduction in vehicle speeds.

Consultation: Local residents through questionnaire and meetings, school, bus company, City Council, emergency services.

Monitoring	Accidents (pia)	Speeds (mph)	Traffic
Before	11 in 4 years	36	7,000
After	0 in 18 months	23	3,750

Contact: John Mitchell   Tel: 0117 929 9074
Authority: Avon County Council

RESIDENTIAL

Technical Data:

Location Type: Residential area, schools.

Road Type and Speed Limit: Urban unclassified: 30 mph.

Scheme Type: Eight flat top humps (including a humped zebra), and a raised junction plateau.

Length of Scheme in Total: 620 m.

Dimensions: Height: 90 mm (av).
Width: 10 m.
Length: Plateau 6 m. Ramp 1 m.
Ramp gradient: 1:11.
Distance from junctions: Varies.

Materials: Plateau: Hot rolled asphalt.
Ramps: Grey block paving.
Kerbs: Precast concrete.
Street furniture: Exposed aggregate concrete bollards.

Signs: Diags. 557 and 1060.

Lighting: Existing street lighting.

Cost: £40,000.

Contact's Comments: Identified objectives have been achieved, however additional work is required to combat displaced traffic in residential roads. Following comments received by the bus company, the ramps are to be lengthened to 1.5 m to increase passenger comfort.

49 NARROWING, FLAT TOP HUMPS
Bradford, Upper Rushton Road

Location: West Yorkshire: Bradford, Upper Rushton Road.

Implemented: May 1992.

Background: Upper Rushton Road is on the outskirts of Bradford approximately 2.5 km from the City Centre. It is 750 metres in length, subject to a 30 mph speed limit and had sub-standard lighting. It is straight throughout its length which encouraged high speeds, totally inappropriate to its residential nature.

Need for Measures: High incidence of rat running and excessive vehicle speeds which led to a high number of pedestrian accidents.

Measures Installed: 7 single way working humps, 2 junction entry humps and 1 junction table.

Special Features: Scheme sympathetic with the local environment. Tactile paving at all features.

Consultation: With police, emergency services, local residents and Passenger Transport Executive.

Monitoring	Accidents (pia)	Speeds (mph)	Traffic (vpd)
Before	13 in 3 years	33.7	3,380
After	0 in 30 months	24.9	2,012

Transportation and Planning
Directorate of Community & Environment
City of Bradford Metropolitan Council

Contact: J M Wallis   Tel: 01274 757401
Authority: City of Bradford Metropolitan Council

Technical Data:

127

Location Type: Urban-residential area.

Road Type and Speed Limit: Urban unclassified. Part mini-bus route. 30 mph.

Scheme Type: Flat top humps with width restrictions, junction entry treatments and junction table.

Length of Scheme in Total: 750 m.

Dimensions: Height 100 mm.
Width: 2.8 m.
Length: Plateau 4.0 m. Ramp 1.2 m.
Ramp gradient 1:12.
Distance from junctions: Average 75 m.

Materials: Plateau and ramps: Buff block paving. Bitumen macadam.
Kerbs: Concrete.
Street furniture: Concrete bollards coated in reflective material.

Signs: 'Road narrows' sign Diag. 516.
Road Markings: Give way (both sides of each treatment). Centre line marking throughout.

Lighting: New 8 m steel columns.

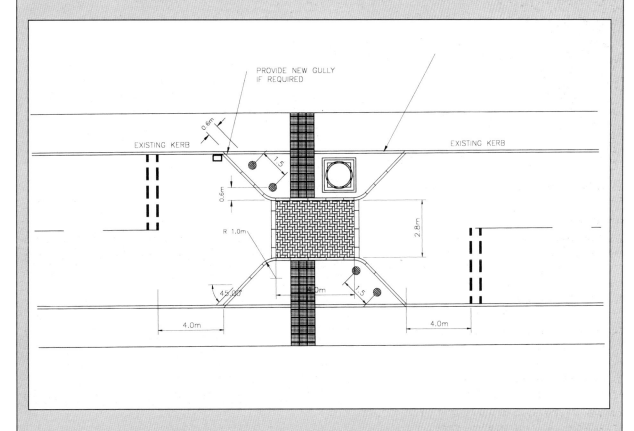

Cost: Department of the Environment Urban Programme Grant (100%) £69,600.

Contact's Comments: None supplied.

50 ROAD HUMP SCHEME
West Wickham, Hawes Lane Area

Location: London: West Wickham, Hawes Lane, Links Road, Rose Walk and Silver Lane.

Implemented: Between February 1991 and February 1993.

Background: These roads are on the periphery of West Wickham High Street.They are some 2.1 km in total length and subject to a 30 mph speed limit with street lighting. They are used as a cut through to avoid the busy junction between the A232 (primary route) and the A214. The area served by these roads contains infant, junior and special schools together with a Health Authority clinic.

Need for Measures: Speeds in Hawes Lane and Links Road led to personal injury accidents and rat running. Road humps were installed in Silver Lane, Rose Walk and the remainder of Hawes Lane to deal with displacement.

Measures Installed: Round top humps.

Special Features: None.

Consultation: Local residents, police and schools. Street notices were also displayed.

Monitoring	Accidents (pia)	Speeds (mph)	Traffic (peak hour)
Before	8 in 3 years	39	694
After	3 in 2 years	21	528

Bromley
THE LONDON BOROUGH

Contact: Dave Chilvery   Tel: 0181 313 4543
Andrew Bashford   Tel: 0181 313 4425
Authority: London Borough of Bromley.

Technical Data:

Location Type: Urban residential.

Road Type and Speed Limit: Urban unclassified: 30 mph.

Scheme Type: Round top humps with tapered sides.

Length of Scheme in Total: Initial scheme1.2 km (18 humps).
Subsequent schemes 0.9 km (13 humps).

Dimensions: Height: 100 mm and 75 mm.
Width: Varies.
Length: 3.7 m.
Ramp gradient: N/A.
Distance from junctions: N/A.

Materials: Bitumen macadam.

Signs: Road hump signs: Diags. 557.1/.2/.3/.4.
Road markings: Diag.1060.1 (Road Hump Markings S.I. 1990 No. 704).

Lighting: Existing.

Cost: London Borough of Bromley Minor Traffic Management budget £31,500.

Contact's Comments: The monitoring information given is for the Hawes Lane/Links Road scheme as this was installed to reduce accidents, whereas subsequent schemes were required to deal with displaced traffic.

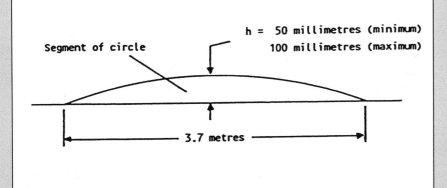

TRANSVERSE SECTION

h = 50 millimetres (minimum)
100 millimetres (maximum)

* 150 millimetres (minimum)
300 millimetres (maximum)

200 millimetres   200 millimetres

LONGITUDINAL SECTION

h = 50 millimetres (minimum)
100 millimetres (maximum)

Segment of circle

3.7 metres

51 CHICANES, GATEWAYS
Tillicoultry, Stalker Avenue

Location: Clackmannan: Tillicoultry, Stalker Avenue.

Implemented: July 1991.

Background: Stalker Avenue is a residential road bounded by a school over part of its length. The road is used as a busy route to the main A91 and is also a bus route.

Need for Measures: High vehicle speeds and a high number of accidents involving young children.

Measures Installed: Single-way 'gates' at either end of the scheme and single-way 'hard chicanes'. The chicanes were designed by Central Regional Council after considerable consultation with various groups, particularly the emergency services and bus companies.

Special Features: Over-run areas for ease of use by buses.

Consultation: A letter was sent to the school, all residents, bus companies, public utilities as well as detailed discussions with the emergency services.

Monitoring	Accidents(pia)	Speeds (mph)	Traffic
Before	10 in 12 years	30+	N/A
After	0 in 3 years	21 (85%ile)	N/A

Contact: Douglas Lawson   Tel: 01786 442729
Authority: Central Regional Council

Central Regional Council

Technical Data:

Location Type: Residential.

Road Type and Speed Limit: Urban unclassified: 30 mph.

Scheme Type: Single way gates and single way hard chicanes.

Length of Scheme in Total: 300 m.

Dimensions: Height: 60 mm overrun.
Width: 0.5 x width of the road.
Length: Island 3.5 m. Overall 15 m.

Materials: 100 mm upstand nibs in standard bituminous construction.
60 mm upstand nibs in concrete block paving.

Signs: Warning signs: Diags. 516, 556 and 570.

Lighting: Existing street lighting was satisfactory.

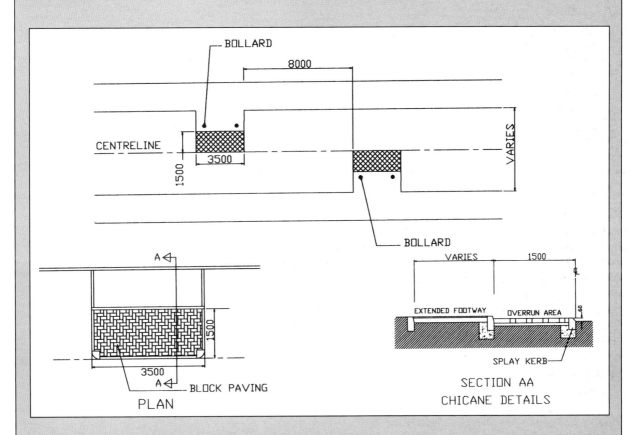

Costs: Central Regional Council Capital budget £10,000.

Contact's Comments: A very successful scheme in reducing vehicle speeds.

52 NARROWING, REFUGE
Barnstaple, Park School

Location: Devon: Barnstaple, Park School, Bishops Tawton Road.

Implemented: September 1991.

Background: This is a major crossing site on a 'B' class road with the main entrance of an adjacent secondary school and the junction of a footway/cycleway. It is a busy commuter route on the edge of town. The road approach from the south was wide with the site at the top of a crest.

Need for Measures: To slow traffic and make provision for a safer crossing point for pedestrians and cyclists.

Measures Installed: Central refuge, horizontal lane deviation with smaller approach islands and hatched central area road markings continued from right lane turn.

Special Features: Brick crossing on carriageway and white lining to emphasize narrowing.

Consultation: Town and District councils, utilities and local secondary school.

Monitoring	Accidents (pia)	Speeds (mph)	Traffic (16 hr)
Before	1 in 5 years	40	7,000
After	0 in 2 years	25	7,000

DEVON COUNTY COUNCIL

Contact: David Netherway Tel: 01271 388503

Authority: Devon County Council

Technical Data:

Location Type: Residential, school, commuter route.

Road Type and Speed Limit: Urban 'B' class road: 30 mph.

Scheme Type: Narrowing of approach to central refuge and crossing point.

Length of Scheme in Total: 250 m.

Dimensions: Width: Refuge 2 m. Lane 3m.
Length: Scheme 250 m. Deviation 56 m.

Materials: Crossing: Red concrete block paving.
Kerbs: Precast concrete.
Islands: Granite setts surface, illuminated bollards and high visibility poles.

Signs: Regulatory signs: Diag. 545 with 546 plate. Diag. 544.3 'Cycle route ahead'.
Road Markings: Diags. 1040,1038 and 1004.

Lighting: Illuminated beacon on central refuge.

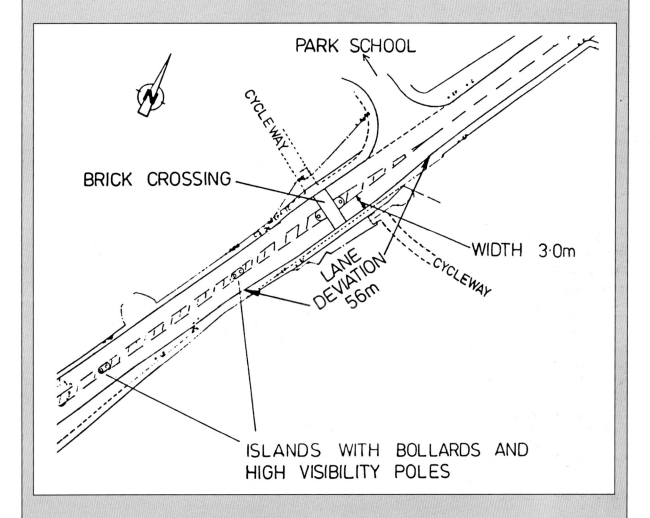

Cost: Devon County Council £20,000.

Contact's Comments: Vehicle speeds very much reduced by horizontal lane deviation. Further approach islands may be added to extend the scheme. Brick crossing construction must be designed to channelise traffic.

53 NARROWING , FLAT TOP HUMPS
Exeter, Burnthouse Lane

Location: Devon: Exeter, Burnthouse Lane.

Implemented: September 1988.

Background: Burnthouse Lane is a residential street on the eastern outskirts of Exeter. Prior to the scheme being implemented, the carriageway was 12.5 metres wide. This, combined with the straight alignment, encouraged high speeds and created hazards for pedestrians.

Need for Measures: To reduce traffic speeds and improve road safety.

Measures Installed: The carriageway was narrowed to 5.5 m, with 1 m cycle lanes and 2.5 m sheltered parking bays on both sides of the road. Concrete block road humps were constructed, and at intervals along the route lateral displacements were introduced to further control speeds.

Special Features: Spherical lighting units, tree planting and raised brick flower beds.

Consultation: A 'Community Liaison Group' was set up for this scheme, including local members, the Community Association, school heads, the emergency services, the bus company and other local representatives.

Monitoring	Accidents (pia)	Speeds (mph)	Traffic (16 hr)
Before	33 in 5 years	34	6,200
After	7 in 5 years	24	5,500

DEVON COUNTY COUNCIL

Contact: Richard Oldfield   Tel: 01392 383800
Authority: Devon County Council.

Technical Data:

135

Location Type: Residential and shopping area.

Road Type and Speed Limit: Urban unclassified: 30 mph.

Scheme Type: Narrowings and flat top humps.

Length of Scheme in Total: 600 m.

Dimensions: Height: 85 mm.
Width: 7.5 m.
Length: Plateau 4 m. Ramp 1 m.
Ramp gradient: 1:12.

Materials: Plateaus: Brindle concrete blocks.
Ramps: Bitumen macadam.

Signs: Road hump warning signs: Diag. 557.

Lighting: Long outreach columns at back of footway.
Feature lighting units at the road humps.

Cost: Devon County Council £220,000.

Contact's Comments: The scheme has been successful in reducing both traffic speeds and accident numbers. It was a finalist in the 1993 *Urban Street Environment* Traffic Calming Award competition.

NARROWINGS, FLAT TOP HUMPS, MINIS
Exmouth, Withycombe Village Rd

Location: Devon: Exmouth, Withycombe Village Road.

Implemented: June 1991.

Background: Withycombe Village Road lies within a primarily residential area, but also serves two schools and a local shopping centre. The section treated is about 500 metres in length.

Need for Measures: To reduce traffic speed and improve road safety.

Measures Installed: Concrete block flat top road humps were installed, with mini roundabouts at each end to act as speed reducing features. The opportunity was also taken to provide sheltered parking on the south side of the road, and a new section of footway.

Special Features: Tree planting on the footway extensions.

Consultation: Local members, school headteachers and the emergency services.

Monitoring	Accidents (pia)	Speeds (mph)	Traffic (16 hr)
Before	14 in 3 years	33	7,900
After	1 in 3 years	25	7,000

Contact: Richard Oldfield   Tel: 01392 383800
Authority: Devon County Council

Technical Data:

137

Location Type: Residential and shopping area.

Road Type and Speed Limit: Urban class 'C' road: 30 mph.

Scheme Type: Narrowings and flat top humps.

Length of Scheme in Total: 500 m.

Dimensions: Height: 85 mm.
Width: 6 m.
Length: Plateau 4 m. Ramp 1 m.
Ramp gradient: 1:12.

Materials: Plateaus: Brindle concrete blocks.
Ramps: Bitumen macadam.

Signs: Road hump warning signs: Diag. 557.

Lighting: Existing lighting not replaced.

Cost: Devon County Council £90,000.

East Devon District Council £1,000 for landscaping.

Total £91,000.

Contact's Comments: The scheme has been successful in reducing both traffic speeds and accident numbers.

55 CHICANES, REFUGES, PROTECTED PARKING, MINI
Plymouth, Beaumont Road

Location: Devon: Plymouth, Beaumont Road.

Implemented: May, 1991.

Background: Beaumont Road is a wide straight length of road through a residential area and is used as a rat-run to and from the city centre. It has street lighting and is subject to a 30 mph speed limit. The treated length is 1,100 m.

Need for Measures: To reduce the number of accidents, along with the concern of residents following several serious accidents involving high speeds.

Measures Installed: The carriageway was narrowed to 6.0m creating protected parking bays varying in width from 1.8m to 3.0m. 10 chicanes with central pedestrian refuges were installed along the treated length to reduce speeds and improve safety for pedestrians. The scheme is introduced at its western end by a mini roundabout and at its eastern end by an existing sharp bend.

Special Features: Brick planters were provided at the wider eastern end.

Consultation: Police, emergency services and bus companies

Monitoring	Accidents (pia)	Speeds (mph)	Traffic (10 hr)
Before	16 in 3 years	37	3,900
After	9 in 3 years	32	3,300

Contact: Geoff Barnes   Tel: 01752 385226

Authority: Devon County Council

Technical Data:

139

Location Type: Residential.

Road Type and Speed Limit: Minor road network local distributor: 30 mph.

Scheme Type: Chicanes, refuges.

Length of Scheme in Total: 1.1 km.

Dimensions: Central islands approximately 5 m in length, 1.2 m wide.
Angle of island varies with road width.
Lane width: 3.1 m to 3.2 m at chicanes, 3 m on straight.

Materials: Kerbing: Precast concrete.
Pavers: Concrete block, buff.
Planters: Brick - Blockley X-mixture wire-cut textured.

Signs: Mini roundabout signed.
Otherwise only temporary signing of chicanes.

Lighting: No changes made to existing street lighting.

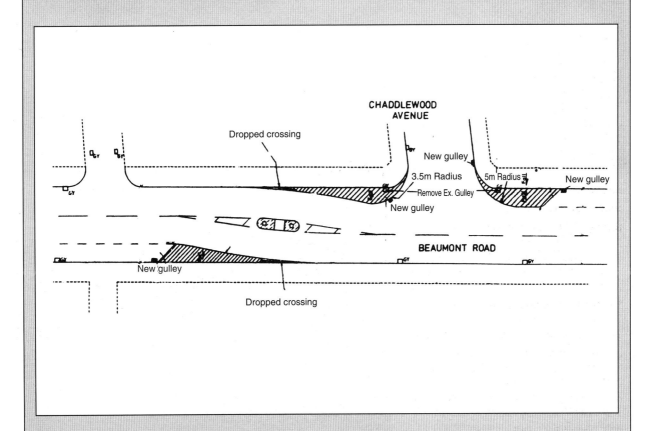

Cost: Devon County Council £99,000.

Contact's Comments: The scheme has proved successful in removing the very high speeds along this length of road and reducing accidents with only a small diversion of traffic into adjacent roads.

56 ROAD MARKING ON DUAL CARRIAGEWAY
Plymouth, Crownhill Road

Location: Devon: Plymouth, Crownhill Road.

Implemented: June 1992.

Background: Crownhill Road is a 3 km length of dual carriageway (formerly part of the A38 Trunk Road through Plymouth) with substandard carriageway width (6.0 m), narrow central reserve with numerous gaps and side junctions, and is mainly subject to a 40 mph speed limit.

Need for Measures: A high accident record occasionally involving high speeds.

Measures Installed: Carriageways narrowed to a single lane in each direction by road markings.

Special Features: Cycle lanes included over full length.

Consultation: Police.

Monitoring (pia)	Accidents (mph)	Speeds	Traffic (10 hr)
Before	82 in 3 years	45	14,000
After	27 in 2 years	41	14,000

DEVON
COUNTY COUNCIL

Contacts: Geoff Barnes   Tel: 01752 385226
Authority: Devon County Council

Technical Data:

Location Type: Residential.

Road Type and Speed Limit: Major road network secondary county route:
40 mph for 2 km, 30 mph for 1 km.

Scheme Type: Road markings.

Length of Scheme in Total: 3 km.

Dimensions: Cycle lanes 1.3 m wide.
Hatching with marked right turn lanes.

Materials: Thermoplastic white paint.

Signs: No changes.

Lighting: No changes made to existing street lighting.

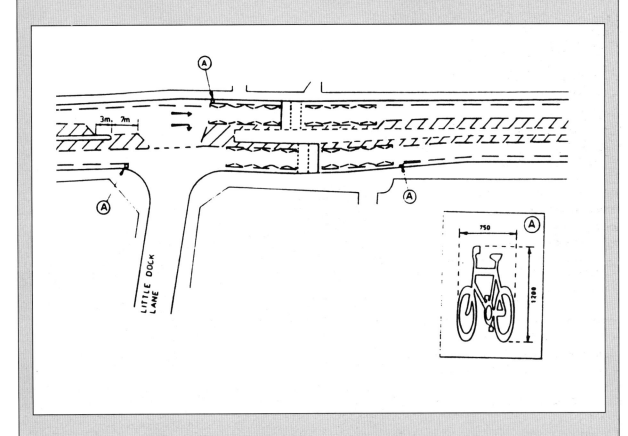

Cost: Devon County Council £12,600.

Contact's Comments: The scheme has created a more controlled traffic flow in Crownhill Road and reduced the very high speeds. It has also shown a significant reduction in the number of personal injury accidents. Cycle facilities have been greatly improved.

57 ROAD HUMPS, REFUGES, NARROWING, JUNCTION PRIORITY CHANGE
Dumfries, Calside Road

Location: Dumfries and Galloway: Dumfries, Calside Road.

Implemented: April 1993.

Background: Calside Road is a distributor road within a large housing estate and is some 1,100 metres in length. Along the frontage of the scheme there are two primary schools (one with a school crossing patrol operating on Calside Road), a playing park, shops and a public house.

Need for Measures: To reduce vehicle speeds, improve driver discipline and improve road safety for pedestrians.

Measures Installed: Seven flat top humps, one round top hump, an alteration to the horizontal alignment, localised road narrowing, provision of a protected parking area, a change of junction priority, two pedestrian refuges and additional waiting restrictions.

Special Features: Pedestrian refuges at the main entry points to the scheme.

Consultation: Community Council, schools, school boards, Parent Teacher Associations, emergency services, bus company and residents, by means of a publicised display.

Monitoring	Accidents (pia)	Speeds (85%ile mph)	Traffic (24 hr)	
			north end	south end
Before	1 in 3 years	35	5,000	3,000
After	0 in 1 year	25	5,000	3,000

Dumfries and Galloway Regional Council

Contact: John Nelson Tel: 01387 60141
Authority: Dumfries and Galloway Regional Council

Technical Data:

Location Type: Residential, schools, shops, public park.

Road Type and Speed Limit: Urban unclassified: 30 mph.

Type of Scheme: Flat top humps, a round top hump, an alteration to the horizontal alignment, provision of protected parking, a change of junction priority and pedestrian refuges.

Length of Scheme in Total: 725 m.

Dimensions	Flat top humps	Round top humps
Height:	100 m	50 mm.
Width:	7.3 m (6). 6 m (1)	7.3 m.
Length:	Plateau 3 m. Ramp 1.2 m.	3.7 m.
Gradient	1:12	N/A.

Materials: Plateau: Brindle coloured block paving.
Ramp: Asphalt.
Bollards: Exposed aggregate.

Signs: Road hump signs: Diag. 557.1.
School warning signs: Diag. 545. Playground subplates: Diag. 547.2.
Bollards on islands: Diag. 610 aspect.
Refuge beacons and waiting restriction plates 'No waiting Mon-Fri 8am-4pm'.

Road Markings: Warning lines Diag. 1004. Edge line at humps Diag. 1012.1.
Hump markings as per regulations, and waiting restrictions.

Lighting: Existing.

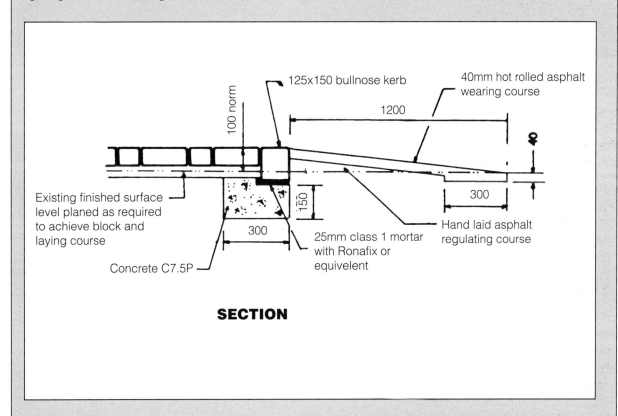

SECTION

Cost: Dumfries and Galloway Regional Council Traffic Calming budget £36,000.

Contact's Comments: The measures have had a marked effect on driver behaviour and vehicle speeds, particularly at school times. The flat humps are used as the preferred locations for children to cross at the frontages of both schools. Some instances have been identified when a driver, approaching from the right, has given way to allow children to cross and children have stepped off the footway expecting drivers approaching from the left to do likewise. The police are further advising schoolchildren on this and advisory signs/markings have been submitted to the Scottish Office for their consideration.

58 NARROWING, FLAT TOP HUMP, MINI
Harlow, Parringdon Road Area

Location: Essex: Harlow, Parringdon Road, Commonside Road, Trotters Road and Parnall Road.

Implemented: February 1991.

Background: The roads in question form a spinal route through a residential area which is also a primary bus route. The scheme was introduced on an area wide basis and was the first of its kind in Essex. The highest recorded speeds were on significant downhill slopes. The measures were designed to reduce the number and severity of personal injury accidents.

Need for Measures: Implemented as an accident remedial scheme following 91 personal injury accidents in three years, one third of which involved pedestrians. Traditional engineering methods such as zebra crossings had failed and complaints of speeding were received from residents.

Measures Installed: 18 ramped narrows with flat top humps and priority working with one mini roundabout.

Consultations: With residents in area, police, emergency services and bus companies. An explanatory leaflet of the final scheme was also dropped to 7,500 householders, all schoolchildren in the nine schools in the area, health centres and community centres.

Monitoring	Accidents (pia)	Speeds (85%ile mph)	Pedestrian Accidents (pia)
Before	30 in 1 year	33	9.6 in 1 year
After	15 in 1 year	27	3.6 in 1 year

Contact: Nicola Foster   Tel: 01245 437113

Authority: Essex County Council

Essex County Council
Highways

Technical Data

Location Type: Urban residential road with high accident rate.

Road Type and Speed Limit: Urban unclassified: 30 mph

Type of Scheme: Ramped narrows with flat top humps and priority working and one mini roundabout.

Length of Scheme in Total: 1 mile square approximately.

Dimensions: Height: 100 mm.
Width: 3.7 m.
Length: approx. 12 m.
Ramp gradient: 1:15.
Distance from junctions: Maximum of 40 m.

Materials: Blocks.

Signs: Diags. 516, 615, 811, and 1061.

Lighting: As required.

Costs: Civils £75,000.
Explanatory leaflets £2,000.
Total £77,000. Funded from Essex County Council capital budget.

Contact's Comments: Brickwork has deteriorated and requires regular maintenance. Central layout caused conflict as drivers, not understanding the priority signs, raced for the gap. Traffic flows have reduced by 15% during the morning 7am-9am peak and by 9% during the evening 4pm-6pm peak. Traffic increase of 1.22% on main route.

59 ROUND TOP HUMPS

Hadleigh, Scrub Lane

Location: Essex: Hadleigh, Scrub Lane.

Implemented: September 1991.

Background: Scrub Lane is situated in the busy south-east of the county. It is a residential road with many junctions and accesses, running parallel to the A13 and is used as a commuter cut-through.

Need for Measures: Speed and volume of traffic inappropriate for a residential location.

Measures Installed: 17 round top road jumps, the number and locations of which were governed by junctions and accesses.

Consultations: With all frontagers, police, emergency services and bus operators.

Technical Data

Location Type: Urban residential road used as commuter cut through.

Road Type and Speed Limit: Urban unclassified: 30 mph.

Type of Scheme: Round topped humps.

Length of Scheme in Total: 1 km.

Dimensions: Height: 50 mm (4), 100mm (13). Width: 6 m. Length: 3.7 m. Ramp gradient: 1:15. Distance from junctions: Maximum of 40 m.

Materials: Hot rolled gravel. Asphalt base course, maximum 85 mm thickness. 6 mm DBM. Wearing course 15 mm thickness.

Signs: 'Road humps' signs to Diags. 557.1, 557.2, 557.3 and 557.4. Road markings to Diag. 1061.1

Lighting: Seven new 5 m lamp columns were required.

Costs: Civils £23,000. Initial experiment £3,000. Monitoring £20,000. Funded from Essex County Council Capital Budget. Total £46,000.

Contact's Comments: The scheme was rushed in because of political interest but more 'before' monitoring should have been done to inform response to public concerns. Residents love the scheme but those in surrounding roads perceive traffic migration. Subsequent monitoring shows this has not happened.

The L10 Noise level (exceeded for 10% of the relevant time period) was reduced by 3.2dB(A). The average noise reduction of a single vehicle pass was 6.6dB(A) but the maximum instantaneous noise level increased by 2.2dB(A). An attitude survey was conducted in Scrub Lane and in the surrounding roads, both on a door to door basis and as a drivers' survey. It was evident that attitudes differed depending on where the interviewee lived.

Monitoring	Accidents (pia)	Speeds (85%ile mph)	Traffic (24 hr)
Before	3 in 11 months	40	8,700
After	0 in 11 months	27	3,580

Contact: Nicola Foster Tel: 01245 437113
Authority: Essex County Council

Essex County Council
Highways

60 RUMBLE STRIPS
RESIDENTIAL
Bexley, Christchurch Road Area

Location: London: Bexley, Christchurch Road, Glenmore Road, New Road, Palmeira Road and Westbrooke Road, Wendover Way.

Implemented: Late August/early September 1989.

Background: The roads selected were all designated as 'other local roads' in the Borough Plan. They were chosen on the basis of complaints and requests from residents that something be done to slow vehicles down. No information was available about the effects of rumble strips and hence an experiment to assess their value was carried out.

Need for Measures: An assessment of need in terms of accident rates was not carried out as the measures were both low cost and experimental.

Measures Installed: Thermoplastic 'rumble strips' laid in sets of 5 or 6 in a variety of configurations.

Special Features: None

Consultation: Before the strips were laid a letter was sent to residents of each road asking for their views on the strips once they were in place. Many letters of support were received before the strips were laid. However, shortly after installation, many complaints of noise and ineffectiveness were received. A second consultation showed that the majority of residents wanted the strips removed. Cyclists and the disabled found the strips uncomfortable.

Technical Data:

Location Type: Suburban residential area.

Road Type and Speed Limit: Local roads, generally only suitable for providing access to land and buildings in the immediate vicinity.

Scheme Type: Thermoplastic rumble strips.

Dimensions: Height: 18 mm. Width: 150 mm.
Spacing: Sets of 5 or 6 strips at 30-40 metre intervals. Interval between each strip within a set 2.5 m or 3.9 m. The strips were stopped short of the kerb so that surface water drainage was not impeded.

Materials: Thermoplastic rumble strips - colour orange.

Signs: None.

Lighting: Existing street lighting.

Cost: Total £9,300. Average £1,860 per street.

Contact's Comments: Although initially, the strips were effective in slowing down traffic after a period of time speeds rose almost to their original levels. All the strips were removed by early 1990. This type of measure may be more suited to areas not fronted by residential properties.

Monitoring	Speed Before (85%ile mph)	Speed After (85%ile mph)
Christchurch Road	36.4	31.4
Glenmore Road	39.6	35.2
New Road	45.2	38.9
Palmeria Road	37.6	35.0
Wendover Road	34.3	28.7

Contact: Alan Stirling   Tel: 0181 303 7777 ext. 3613
Authority: Bexley London Borough

61 ROAD CLOSURES
Gloucester, Conduit Street

Location: Gloucestershire: Gloucester, Tredworth, Conduit Street.

Implementation Date: September 1992.

Background: Conduit Street, 545 metres in length and bounded by inner City residential housing, acted as a feeder road for seven other roads or streets around it.

Need for Measures: To reduce the crime and traffic related rat-run through the maze of local streets.

Measures Installed: Closure of four roads at their junctions with Conduit Street by means of footway extensions.

Special Features: Drop kerbs to form appropriate crossing points for pedestrians and cyclists, together with suitably spaced metal bollards, and kerb alterations for vehicle turning areas.

Consultation: Local residents, Police and other emergency services.

Monitoring: Information not available.

GLOUCESTER CITY COUNCIL

Contact: Clive Ashby   Tel: 01452 396798
Authority: Gloucester City Council

Technical Data:

Location Type: Residential.

Road Type and Speed limit: Urban unclassified: 30 mph.

Scheme Type: Closure of side roads at junctions with local feeder road.

Length of Scheme in Total: 545 m.

Dimensions: Height: Bollards 1 m height, spacing 1.2 m (max).
Width: Footway 1.75 m average.
Length: 5 m overall per road closure.
Ramp gradient: Normal for drop kerbs.
Distance from Junctions: N/A.

Materials: Footway: Bitmac surfacing on stone base.
Kerbs: Precast concrete.
Street Furniture: Concrete filled steel tube bollards.

Signs: 'No through Road', street name plates and home number plates giving
access information along each closed-off section.

Lighting: No alterations to existing.

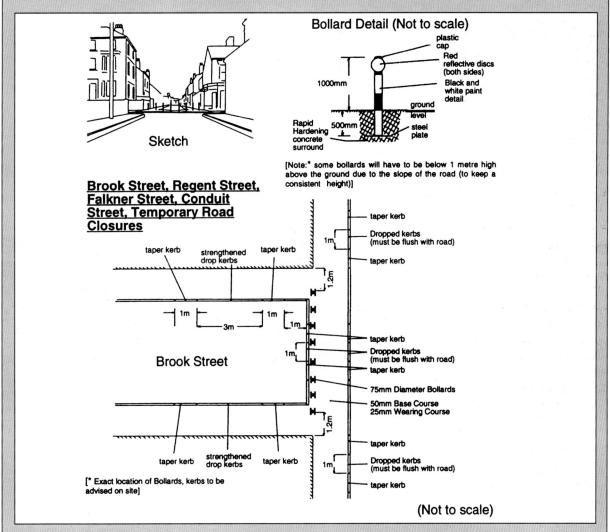

Sketch

Bollard Detail (Not to scale)

plastic cap
Red reflective discs (both sides)
1000mm
Black and white paint detail
ground level
Rapid Hardening concrete surround
500mm
steel plate

[Note:* some bollards will have to be below 1 metre high above the ground due to the slope of the road (to keep a consistent height)]

Brook Street, Regent Street, Falkner Street, Conduit Street, Temporary Road Closures

taper kerb   strengthened drop kerbs   taper kerb

1m   3m   1m

1m

1m

Brook Street

1.2m

taper kerb
Dropped kerbs (must be flush with road)
taper kerb

taper kerb
Dropped kerbs (must be flush with road)
taper kerb
75mm Diameter Bollards
50mm Base Course
25mm Wearing Course

1.2m

taper kerb   strengthened drop kerbs   taper kerb

taper kerb
Dropped kerbs (must be flush with road)
taper kerb

[* Exact location of Bollards, kerbs to be advised on site]

(Not to scale)

Cost: £10,500.

Contact's Comments: A good low cost effective scheme which has reduced traffic and improved road safety within the whole of the town's inner area.

62 ONE-WAY STREET, CHICANE, HUMPS, PINCH POINT, EXIT BUILT-OUT
Newport, Llanthewy Road

Location: Gwent: Newport, Llanthewy Road.

Implemented: May 1993.

Background: Llanthewy Road is a residential street on the periphery of the central area of Newport. The road has an overall gradient of 8%, with a maximum value of 10% at its southern end. It falls in a northeasterly direction towards the town centre. There is 'No exit' out of Llanthewy Road at its southern end.

Need for Measures: To reduce the speed and volume of short cutting traffic.

Measures Installed: One way street, entry chicane, 2 round top humps, 1 pinch-point with a flat top hump and an exit build-out.

Special Features: An entry chicane.

Consultation: Public exhibition, emergency services, Borough Council.

Monitoring	Accidents (mph)	Speeds	Traffic (10 hr)
Before	2 in 3 years	32	2,228
After	None to date	20	1,827 one way

GWENT County Council

Contac: Robert Campbell Tel: 01633 832725

Authority: Gwent County Council

Technical Data:

Location Type: Residential, close to town centre.

Road Type and Speed Limit: Urban unclassified: 30 mph.

Scheme Type: Southern section made one way preventing homeward short cutting traffic; this section also has an entry chicane, 2 round top road humps, 1 flat top road hump with a pinch-point to 3.5 metres and an exit build-out.

Length of Scheme in Total: 600 m.

Dimensions: Round top hump: Height 100 mm. Width 8.5 m. Length 3.7 m.

Flat top hump: Height 100 mm. Width 3.5 m. Length 3.7 m.
   Plateau 2.5 m. Ramp gradient 1:6.

Chicane: Overall length 25 m.

Materials: Plateau and ramp: Blockwork paviours.
Round top humps, chicane and build-out: Bitmac.

Signs: 15 signs in total.

Lighting: No change.

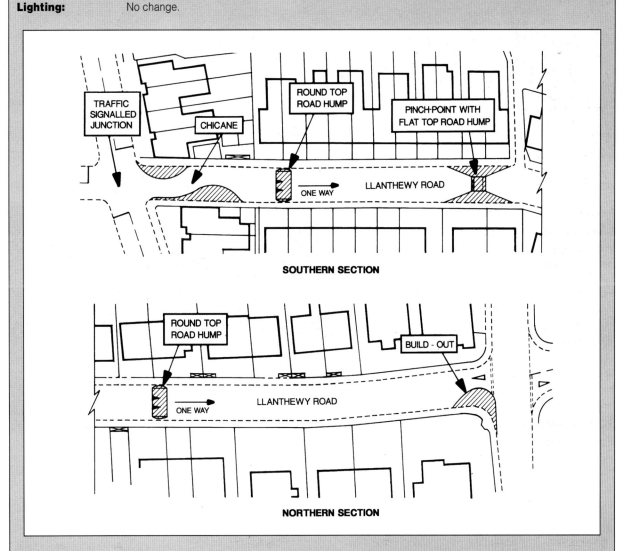

SOUTHERN SECTION

NORTHERN SECTION

Cost: Gwent County Council £25,000.

Contact's Comments: None supplied.

63 PRIORITY NARROWING, MINIS, ROAD CLOSURE, ROAD HUMPS, CYCLE LANE
Hemel Hempstead, Peascroft Road

Location: Hertfordshire: Hemel Hempstead, Peascroft Road.

Implemented: January 1992.

Background: Peascroft Road is situated in the middle of a.large residential area. Two schools front onto Peascroft Road and the road becomes very congested during picking up and dropping off times. The road is 850 metres in length and subject to a 30 mph speed limit.

Need for Measures: High speed accidents and rat running.

Measures Installed: Round top humps and kerb to kerb block work flat top humps, mini roundabouts, road closure and a humped priority narrowing with cycle lane.

Special Features: Vandalproof base-lit bollards.

Consultation: District Council and members, County Council members, emergency services, police, bus company, public and utilities.

Monitoring	Accidents (pia)	Speeds (85%ile mph)	Traffic Flow (am peak)
Before	17 in 5 years	44	723
After	1 in 2 years	18	265

Hertfordshire
COUNTY COUNCIL
Transportation

Contact: David Bowie   Tel: 01992 556148
Authority: Hertfordshire County Council

Technical Data:

153

Location Type: Residential, schools.

Road Type and Speed Limit: Urban unclassified: 30 mph.

Scheme Type: Round top and flat top humps with mini roundabouts and a road closure.

Length of Scheme in Total: 850 m.

Dimensions: Height: Flat top humps 100 mm. Round top humps 75 mm.
Width: Varies with width of road.
Length: Round top humps 3.7 m. Flat top: Plateau 2.5 m. Ramp 1.2 m.
Ramp gradient: 1:12.
Distance from junctions: No greater than 40 m.

Materials: Plateau: Brindle brick paving held in position by two rows of concrete channels
Ramps: hot rolled asphalt.
Kerbs and channels: Precast concrete
Street furniture: Waisted cast iron bollards. Hazard marker posts with Class 1 reflective faces.

Signs: Diags. 557.1 and 557.3.

Lighting: Uprated to BS5489

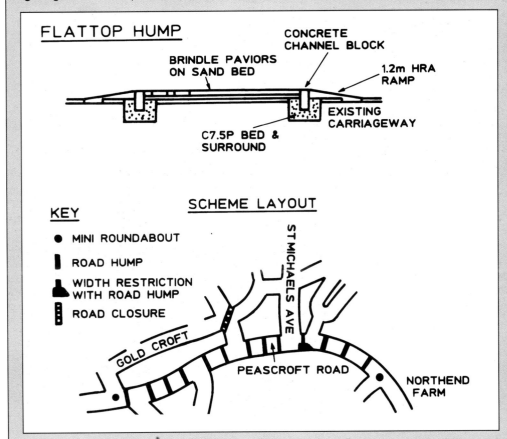

Cost: Civils £36,000.
Street Lighting £16,000.
Total £52,000, from Traffic Supplementary Grant.

Contact's Comments: A significant reduction in both speed and non speed related injury accidents.
Scheme welcomed by local residents and adjacent schools.

64 SPEED TABLE, CHICANES, PINCH POINTS
Shepherds Bush, Wulfstan St

Location: London: Shepherds Bush, Wulfstan Street.

Implementation Date: October-December 1989.

Background: Wulfstan Street is situated within a popular cottage style terraced housing estate where a close knit family orientated community lives. The tree lined street is 700 metres long, straight, well lit, with a 30 mph speed limit, and is heavily parked. The street is used as a car commuter cut-through which divides the community from local facilities such as the local primary school, hospital, park and shops.

Need for Measures: Parked cars, fast traffic, and many small children crossing created a real and increasing accident problem in the street which needed tackling.

Measures installed: Speed table, chicanes, pinch points, wide pedestrian islands and side road entry treatments.

Special Features: Measures designed to integrate with the symmetry of the unusual road layout and housing design in the conservation area.

Consultation: Meetings were held with the local community group and the emergency services. No bus or cycle routes were affected.

Monitoring	Accidents (pia)	Speeds (mph)	Traffic Flow (vpd)
Before	14 in 3 years	35	820
After	2 in 3 years	20	445

Contact: Christopher Boylan   Tel: 0181 748 3020 Ext. 3344

Authority: London Borough of Hammersmith & Fulham

Technical Data:

Location Type: Urban residential. Two-storey terraced houses. Conservation area status.

Road Type and Speed Limit: Urban unclassified: 30 mph.

Type of Scheme: One speed table, two junction chicanes, two pinch points, three wide pedestrian islands, three side road entry treatments.

Total Length of Scheme: 700 m.

Materials: Various including the use of plastic reflective bollards.

Signs: Diags. 516, 517, 602, 615 and 811.

Markings: hatching, centre, give way.

Lighting: Repositioned lamps to highlight kerb build-outs.

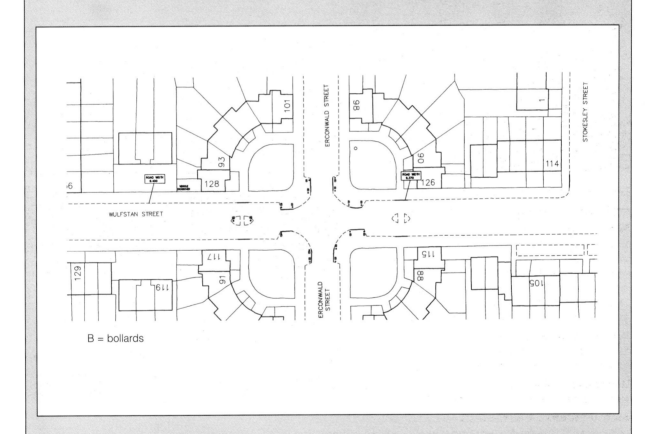

B = bollards

Cost: 1989/90 Transport Policies and Programme funds £70,000.

Contact's Comments: The scheme has maintained the accident saving since 1989 and outperformed expectations.

65 ENTRY TREATMENT, FLAT TOP HUMPS, MINI
Grimsby, Grant Thorold Park

Location: South Humberside: Grimsby, Grant Thorold Park area.

Implemented: May 1992.

Background: Grant Thorold forms part of a large zone within Grimsby which was considered suitable for area-wide accident remedial measures. The scheme covers an area of approximately 1 sq km and is mainly residential in character, with a few local shops.

Need for Measures: To inhibit crossover movements into and out of the area by 'rat running' traffic, and so reduce accidents at junctions.

Measures Installed: 8 raised junction entry treatments, 3 speed tables, 1 mini roundabout.

Special Features: Built out areas at entry treatments planted with trees to accord with local environment and, once established, to reduce the optical width thus helping to influence speed

Consultation: Emergency services, Borough Council and residents, through exhibition in local library.

Monitoring	Accidents (pia)	Speeds	Traffic
Before	50 in 3 years	not recorded	not recorded
After	15 in 2 years	not recorded	not recorded

HUMBERSIDE
COUNTY COUNCIL
SERVING THE PEOPLE OF
East Yorkshire & North Lincolnshire

Contact: Eric Wragg Tel: 01482 884032
Authority: Humberside County Council

RESIDENTIAL

Technical Data:

Location Type: Residential, local shops.

Road type and Speed Limit: Urban unclassified: 30 mph.

Scheme Type: Area wide traffic calming, with ramped junctions, speed tables and mini-roundabout.

Area of Scheme in Total: 1 sq km.

Dimensions: Height: l00 mm.
Width: 6.5 m.
Length: Plateau 2.5 m. Ramps 2 m.
Ramp gradient: 1:20.

Materials: Plateaus and ramps: Red block paving.
Carriageway narrowing at junctions: Grey block paving.
Kerbs: Precast concrete.
Street furniture: Cast iron bollards with reflective material at top.

Signs: Diags. 562 and 563 at speed tables.
Non prescribed 'Road Narrows' and 'Ramps Ahead' signs at the entry points to the zone.

Lighting: 6 m group B, low pressure sodium.

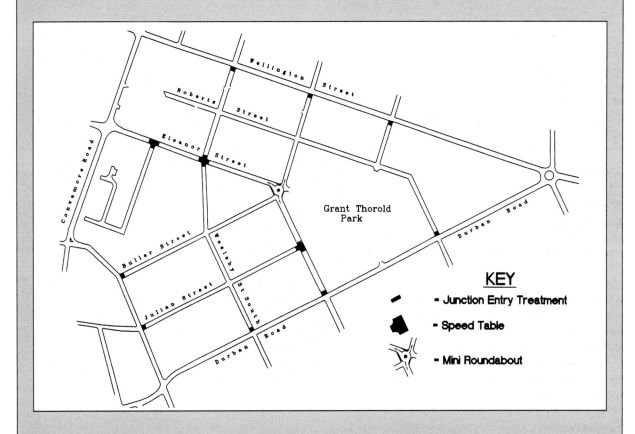

Cost: Humberside County Council Local Safety Schemes budget £122,422.

Contact's Comments: The scheme appears to have been successful in reducing accidents, with a 45% reduction from the before period. With hindsight the number of signs indicating the humps could have been reduced to lessen the effect of sign clutter in Eleanor Street.

66 CHICANES, PROTECTED PARKING, REFUGES
Grimsby, Nunsthorpe Estate

Location: South Humberside: Grimsby, Nunsthorpe Estate, Second Avenue.

Implemented: May 1991.

Background: The Nunsthorpe area of Grimsby is a large local authority housing estate. The whole area has been examined with a view to implementing accident remedial measures, given the high numbers of accidents involving children. Second Avenue forms an access road to a hospital and is mainly residential with a shopping area to the northern end of its 700 metre length.

Need for Measures: High speeds, accidents involving pedestrians including children, an absence of pedestrian facilities and high demand for on-street parking.

Measures Installed: Built-out kerbs to form sheltered parking bays and chicanes, pedestrian refuge islands and one mini roundabout. A road safety campaign was also mounted in the area.

Special Features: An overrun area was provided near to the shopping area to narrow the carriageway whilst still allowing vehicles to access parking bays without too much difficulty.

Consultation: Emergency services, Borough Council and residents, exhibition in local supermarket.

Monitoring	Accidents (pia)	Speeds	Traffic
Before	16 in 3 years	not recorded	not recorded
After	6 in 3 years	not recorded	not recorded

HUMBERSIDE
COUNTY COUNCIL
SERVING THE PEOPLE OF
East Yorkshire & North Lincolnshire

Contact: Eric Wragg Tel: 01482884032
Authority: Humberside County Council

Technical Data

Location Type: Residential area.

Road type and Speed Limit: Urban unclassified: 30 mph.

Scheme Type: Chicanes, staggered parking, refuge islands and mini roundabout

Length of Scheme in Total: 700 m.

Dimensions: No vertical measures. Built out kerbline to form a chicane at 50 m intervals.

Materials: Built out areas: Precast concrete kerbs with grey block paving.

Refuge islands and overrun area: Red block paving.

Street furniture: Concrete bollards with white thermoplastic and applied ballotini on top section of bollards.

Signs: None.

Lighting: Upgraded to high pressure sodium as part of maintenance scheme.

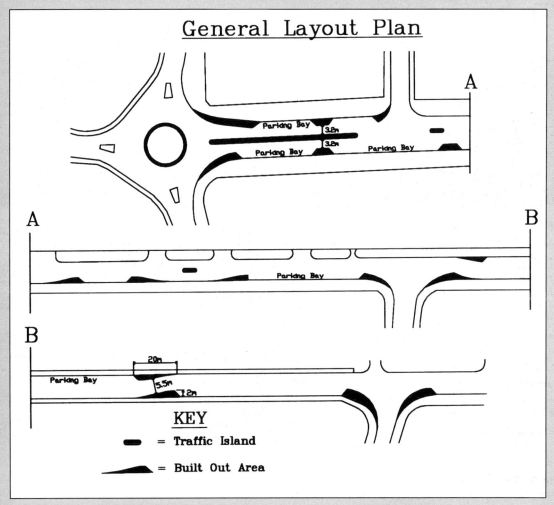

General Layout Plan

KEY

= Traffic Island

= Built Out Area

Cost: Grimsby Borough Council for maintenance work and environmental enhancement £130,000.
Humberside County Council specified maintenance and street lighting budgets £56,000.
Humberside County Council Local Safety Schemes budget.£15,000.
Total £201,000.

Contact's General Comments: The scheme appears to have been successful in reducing accidents, with a 38% reduction from the before period.

67 PLATEAUS, BUS LAY-BYS, NARROWINGS
Newport (IOW), Pan Estate

Location: Isle of Wight: Newport, Pan Estate.

Implemented: June 1993.

Background: Pan Estate is a large housing estate of some 1,100 households, with an above average concentration of school age children. It lies to the east of Newport and has two main commuter routes on its northern and western boundaries, as well as the main Coppins Bridge gyratory system that connects all of the main roads that transit Newport. As a consequence Furrlongs, which is the main route through the centre of Pan Estate, was being used as a 'rat run' and this contributed to the 17 personal injury accidents in a five year period. Pan Estate also has four schools located within it, and consequently attracts a large number of vehicle movements at peak times. The entire area is covered by a 30mph speed limit and a 6'6" width restriction.

Need for Measures: Speed, and rat running traffic that contributed to child pedestrian accidents.

Measures Installed: Numerous flat top tapered edge humps, three 10 m long plateaus, narrowings.

Special Features: Raised plateau incorporating adjacent bus stop layby.

Consultation: Local residents, police, ambulance, fire brigade, schools and bus company.

Monitoring	Accidents (pia)	Speeds (mph)	Traffic (16 hrs)
Before	17 in 5 years	35	4,834
After	5 in 3 years	17	3,182

ISLE of WIGHT
COUNTY COUNCIL

Contact: Mr Peter Taylor Tel: 01983 823763
Authority: Isle of Wight County Council

Technical Data:

Location Type:　　Urban: residential estate.

Road Type and Speed Limit: Urban unclassified: 30 mph.

Scheme Type:　　Narrowing - two way-working with flat top plateaus and tapered edge humps.

Length of Scheme in Total: 610 m.

Dimensions:　　Height: 75 mm.
Width: 5 m (two way working).
Length: 10 m.
Ramp gradient: 1:10.
Distance from junctions: Varies.

Materials:　　Plateaus and ramps: Buff block paving with red ramps and blue/black band between carriageway and bus stop.

Kerbs: Hydraulically pressed pre-cast concrete.

Street furniture: Glasdon 'Admiral' bollards with red reflective bands.

Signs:　　Diags. 611.1, 612.1, 613,3 and 557.1 plus supplementary plates.

Lighting:　　Class A lighting at 35 m centres.

Cost:　　£40,000.

Contact's Comments:　A successful scheme endorsed by a comprehensive public consultation after construction.

68 PLATEAUS, RAISED JUNCTION
Sittingbourne, Stanhope Ave Area

Location: Kent: Sittingbourne, Stanhope Avenue, South Avenue and Chilton Avenue.

Implemented: August 1989.

Background: These roads lie in a residential area south-east of Sittingbourne, and are used by through traffic accessing a major supermarket, and by commuters cutting between the B2163 and A2. They are subject to a 30 mph limit. Stanhope and South Avenues are bus routes and there are two schools in South Avenue.

Need for Measures: Unacceptable level of accidents. There were 2 pedestrian fatalities in the 3 year period prior to 1989. Contributory factors were unreasonably high speeds and excessive volume of traffic. However, no other suitable route for the traffic exists.

Measures Installed: Three two-way KCC style flat top block paved humps with narrowings in each of Stanhope and South Avenues; one similar in Chilton Avenue, together with a table junction.

Consultation: Full public consultation was carried out prior to the scheme being implemented, although attendance at the exhibition was very poor.

Monitoring	Accidents (pia)	Speeds (85%ile mph)	Traffic (12 hr)
Before	12 in 3 years	38 to 40	5,050
After	3 in 3 years	28 to 30	3,500

Contact: Ray Dines   Tel: 01622 696993
Authority: Kent County Council

Kent County Council
HIGHWAYS & TRANSPORTATION

Technical Data:

Location Type: Residential, schools and buses.

Road Type and Speed Limit: Urban unclassified: 30 mph.

Scheme Type: Flat top humps with narrowings, raised junction.

Length of Scheme in Total: 1.2 km.

Dimensions: Height: 100 mm.
Width: 5 m.
Length: 10 m.
Ramp gradient: 1:10.
Distance from junctions: 40 m.

Materials: Plateau and ramps: Concrete blocks - buff.
Street furniture: Verge marked posts - Glaston.

Signs: Ramp: Diag. 516 plated.

Lighting: Columns re-sited at each measure.

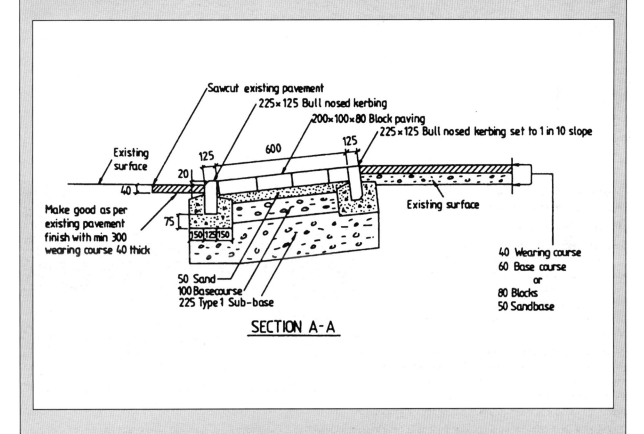

SECTION A-A

Cost: £43,000.

Contact's Comments: Spacing, signing and markings all pre-date current regulations.

69 CHICANES
Tonbridge, Brook Street

Location: Kent: Tonbridge, Brook Street.

Implemented: April 1993.

Background: Brook Street is on the outskirts of Tonbridge. The section considered for traffic calming measures has two schools and a Higher Education College adjoining the road. The road is subject to a 30 mph speed limit and has lighting.

Need for Measures: High speeds, which led to child pedestrian accidents including a fatality.

Measures Installed: A series of pedestrian refuge islands with hatching, permanently marked out parking bays and kerb build-outs to produce chicanes, thus enforcing a 'weaving' movement between the measures.

Consultation: With the police. No residents adjacent to this section of road. No bus route.

Monitoring	Accidents (pia)	Speeds (mph)	Traffic (vpd)
Before	8 in 3 years	44	up to 7,000
After	1 in 1 year	37	up to 6,500

Contact: Lidia Swierczysnka   Tel: 01622 696830
Authority: Kent County Council

Kent County Council
HIGHWAYS & TRANSPORTATION

Technical Data:

165

Location Type: Urban, schools.

Road Type and Speed Limit: Urban classified 'C' road: 30 mph.

Type of Scheme: Chicanes formed by pedestrian refuges and protected parking.

Length of Scheme in Total: 650 m.

Dimensions: Pedestrian refuges: 1.3 m wide.
                         Length of kerb build out: 5 m.

Materials: Kerb build outs: PCC kerbing infilled with HRA.

Signs: Road narrowing: Diags. 516 and 517.

Lighting: Existing columns upgraded and a number of columns re-located.

Cost: £42,000.

Contact's Comments: Initial complaints about students using road as 'race-track' have now died away.

70 PINCH POINTS WITH SPEED CUSHIONS
Huddersfield, Victoria Road

Location: West Yorkshire: Huddersfield, Victoria Road.

Implemented: August 1991.

Background: Victoria Road is a residential road situated 3/4 mile from Huddersfield town centre. Sited along the road is a large primary school and mosque. It was well used as a rat run.

Need for Measures: A high number of young child pedestrian casualties.

Measures Installed: Pinch points and twin 'Berlin' cushions outside the school and mosque. Flat topped rumble strips in advance of pinch points.

Special Features: The pinch points are designed to encourage pedestrians to use as crossing places by using dropped kerbs. Trief has been used to present a formidable appearance to motorists, to emphasise the pinch and deter excessive speeds.

Consultation: All residents invited to a mobile exhibition in the area. The school, local community leaders, emergency services and bus operators.

Monitoring	Accidents (pia)	Speeds (mph)	Traffic Flow (vpd)
Before	21 in 3 years	39	4,000
After	3 in 3 years	24	4,000

Contact: Peter Salmon Tel: 01484 446530
Authority: Kirklees Metropolitan Council

Technical data:

Location Type: Terraced residential, outskirts of town.

Road Type and Speed Limit: Unclassified: 30 mph.

Scheme Type: Pinch point, single way working.

Length of Scheme in total: 650 m.

Dimensions: Pinch point: 3.0 m wide.
Berlin cushions: 1.25 m x 2.3 m x 75 mm high.
Rumble strips:1.0 m wide x 25 mm high at 50 m intervals.

Materials: Pinch point: Trief kerbing.
Berlin cushion: Tegula setts.
Rumble strips: Tegula setts.

Signs: Regulatory: Diag. 811.
Warning: Diag. 562, plus non standard scheme warning signs.

Lighting: No change.

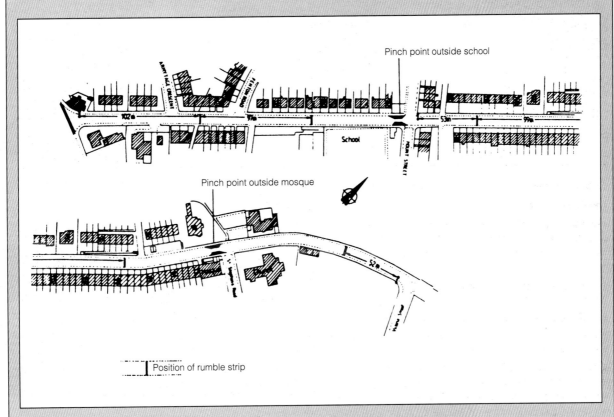

Pinch point outside school

Pinch point outside mosque

School

Position of rumble strip

Cost: £80,000.

Contact's Comments: Successful in reducing accidents by 86% and speeds by 40%. Not successful at reducing bus speeds which were considered to be high.

71 FLAT TOP PLATEAUS, THERMOPLASTIC 'THUMPS'
Golcar, Sycamore Avenue

Location: West Yorkshire: Kirklees, Huddersfield, Golcar, Sycamore Avenue.

Implemented: December 1991.

Background: Sycamore Avenue is the spine road of a predominantly council owned housing estate. At the midpoint of the road, there is a cluster of shops with an access to a large primary school. The road is dead straight and level and subject to a 30 mph limit. It is a bus route and has class B street lighting.

Need for Measures: The number of child pedestrians injured crossing to and from the shops, and a high 85%ile speed.

Measures Installed: Flat top plateaus on each approach to the shopping area. Dummy plateau in the middle. Yellow thermoplastic 'thumps' on remainder of road.

Special Features: Retention of parking outside the shops without prejudicing the safety aspect.

Consultation: All residents invited to a mobile exhibition in the area. Emergency services, bus operators, utilities, etc.

Monitoring	Accidents (pia)	Speeds (mph)	Traffic
Before	9 in 3 years	35	5,000
After	1 in 30 months	22	5,000

Contact: Peter Salmon Tel: 01484 446530

Authority: Kirklees Metropolitan Council

Technical data:

Location Type: Residential estate, school, shops.

Road Type and Speed Limit: Urban unclassified: 30 mph.

Scheme Type: Narrowing of carriageway, provision of parking space, ramped plateaus and thermoplastic 'thumps'.

Length of Scheme in total: 600 m.

Dimensions: Plateaus: Height 100 mm. Width 5.0 m. Length 20.0m (av). Gradient 1:10.

Thermoplastic 'thumps': Height 60 mm (max). Width 900 mm. Length 6.5m. Colour - yellow.

Materials: Plateaus: Red DBM.

Kerbs: Precast concrete.

Street Furniture: Pedestrian guardrail. Concrete planters

Signs: Diag: 562, plate: 'Ramps', plate: 'Drive slowly', on yellow backing board

Lighting: 6.0 m 70 watt HP sodium

Cost: £80,000.

Contact's Comments: The scheme works well, looks good environmentally and is proving effective. But double decker buses have difficulty with the 1:10 gradient ramps.

72 'CROWNED' RUMBLE STRIPS
Paddock, Church Street

Location: West Yorkshire: Kirklees, Huddersfield, Paddock, Church Street.

Implemented: January 1989.

Background: Church Street is an urban distributor serving a large residential suburb. The frontage is mixed commercial and residential.

Need for Measures: The road was identified as a site for concern with an average accident rate four times the national average. Excessive speed was a dominant factor.

Measures Installed: Crowned rumble strips at 50 m intervals.

Special Features: The strips were rounded to a finished height of 35 mm.

Consultation: All residents and traders, emergency services and bus operators were consulted. A very hostile reception was received from traders and bus operators.

Monitoring	Accidents (pia)	Speeds (mph)	Traffic (vpd)
Before	24 in 3 years	35	7,000
After	1 in 2 years	24	5,500

Contact: Peter Salmon Tel: 01484 446530
Authority: Kirklees Metropolitan Council

Technical data:

Location type: Urban mixed commercial and residential use area.

Road Type and Speed Limit: Urban unclassified: 30 mph.

Scheme Type: Rounded rumble strips.

Length of Scheme in total: 900 m.

Dimensions: Height 35 mm.
Width 5.0 m.
Length 1.0 m.
Upstand 15 mm.
Spacing 50 m.

Materials: Stone setts with pre-cast concrete constraints.

Signs: Warning signs to Diag. 562, plate 'Ramps Ahead'.

Lighting: No change.

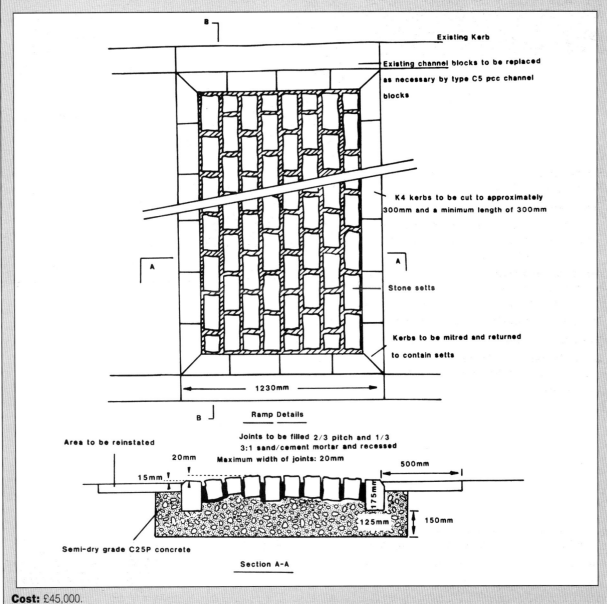

Existing Kerb

Existing channel blocks to be replaced as necessary by type C5 pcc channel blocks

K4 kerbs to be cut to approximately 300mm and a minimum length of 300mm

Stone setts

Kerbs to be mitred and returned to contain setts

1230mm

Ramp Details

Area to be reinstated

Joints to be filled 2/3 pitch and 1/3 3:1 sand/cement mortar and recessed
Maximum width of joints: 20mm

20mm

15mm

500mm

175mm

125mm

150mm

Semi-dry grade C25P concrete

Section A-A

Cost: £45,000.

Contact Comments: Notorious, contentious, but successful in reducing speeds and casualties.

73 PLATEAUS

RESIDENTIAL

Ravensthorpe, North Road

Location: West Yorkshire: Kirklees, Ravensthorpe, North Road.

Implemented: May 1991.

Background: North Road is a well used through route from Ravensthorpe to Mirfield. The road is predominantly residential and sited on the mid-section of the treated length is a large primary school. The road is frequently used by buses.

Need for Measures: A high number of child pedestrian accidents.

Measures Installed: Plateaus.

Special Features: The large number of plateaus used with spacings of 40 m - 50 m to ensure a constant low speed throughout the treated section.

Consultation: All residents were invited to a mobile exhibition in the area. Emergency services and bus operators were formally consulted.

Technical data:

Location type: Urban residential area, school, through route.

Road Type and Speed Limit: Urban unclassified: 30 mph.

Scheme Type: Plateaus with extended kerb lines at the junctions.

Length of Scheme in total: 500 m.

Dimensions: Height 100 mm. Width 6.0 m. Length 8.0 m. Ramp gradient 1:10. Spacings: 40 m - 50 m

Materials: Red asphalt. Pre-cast concrete kerbs.

Signs: Standard road hump markings to Diag. 1061. Special non-prescribed warning signs. Road markings.

Lighting: New 90 watt LP sodium.

Cost: £80,000.

Contact's Comments: Generally well accepted.

Monitoring	Accidents (pia)	Speeds (mph)	Traffic
Before	10 in 3 years	35	8,500
After	3 in 3 years	22	6,500

Contact: Peter Salmon Tel: 01484 446530

Authority: Kirklees Metropolitan Council

74 ROUND TOP HUMPS
Featherstone, The Avenue

RESIDENTIAL

Location: Staffordshire: The Avenue, Featherstone.

Implemented: October 1991.

Background: Featherstone is a village close to Junction 1 of the M54 and within easy commuting distance of the West Midlands conurbation. The Avenue is the main distributor road serving the housing area of the village. It is subject to a 30 mph speed limit and has street lighting.

Need for Measures: The Avenue is straight, encouraging high speeds and runs past shops and a school. This has led to numerous complaints.

Measures Installed: 75 mm round top humps, 3.7 m in length with tapered sides extended across complete width of road. Approx 150 m apart.

Special Features: This was one of Staffordshire's experimental schemes.

Consultation: Local residents, Parish Council, District Council, emergency services and police.

Technical Data:

Location Type: Suburban housing estate.

Road Type and Speed Limit: Semi-urban: 30 mph.

Scheme Type: Two way round top humps.

Length of Scheme in Total: 800 m (including Brook House Lane).

Dimensions: Height: 75 mm. Width: 7.3 m. Length: 3.6 m. Ramp gradient: 1:24 (av).

Materials: Speed humps: Medium temperature asphalt HRA 55/10.

Signs: 'Road hump' signs: Diag. 557.1. Markings Diag. 1060.1 (including on adjacent A460 main road).

Lighting: Existing.

Cost: Staffordshire County Council Traffic Management budget £8,200.

Contact's Comments: None supplied.

Monitoring	Accidents (pia)	Speeds (mph)	Traffic (5 day av 24 hr)
Before	6 in 3 years	39	3,600
After	2 in 20 months	29	2,800

Staffordshire
County Council

Contact: C Oldham   Tel: 01785 276556

Authority: Staffordshire County Council

75 NARROWING, FLAT-TOP HUMPS, ONE-WAY STREETS
Lancaster, Primrose Area

Location: Lancashire: Lancaster, Dale Street, Primrose Street and Prospect Street.

Implemented: June 1993.

Background: The Primrose Housing Action Area is on the periphery of the historic city centre of Lancaster. The roads in this area, with parking on both sides, are used by motorists to avoid using the city centre gyratory system. The roads are subject to a 30 mph speed limit.

Need for measures: To reduce vehicular speeds and the incidence of 'rat runners'.

Measures Installed: Flat-top humps with associated narrowings. One-way streets. 7.5 tonne lorry weight restriction order.

Special Features: Materials to match the character of the area.

Consultation: With residents, the police, Lancashire County Council, emergency services and bus companies.

Monitoring	Accidents (pia)	Speeds (mph)	Traffic (16 hr)	
			Dale St	Prospect St
Before	9 in 4 years	34	2,600	1,500
After	Not available	15	Not available	

Contact: Jim Robson   Tel: 01524 582604

Authority: Lancaster City Council

Technical Data:

Location Type: Residential ,older housing adjacent tohistoric city centre.

Road Type and Speed Limit: Urban unclassified: 30 mph.

Scheme Type: Flat top humps with associated narrowings. One-way streets.

Length of Scheme in Total: 640 metres on three streets.

Dimensions: Height: 100 mm.

Width: 3 m.

Length:　　Plateau 2.5 m
　　　　　　Entry ramp 1 m.　　Gradient 1:10.
　　　　　　Exit ramp 2 m.　　Gradient 1:20.

Materials: Humps: Buff block paviours.
Kerbs: Recycled natural stone
Footway paving: Natural stone flags.
Bollards: Cast iron.

Signs: Signing and road markings as per regulations.

Lighting: Existing, which had been recently updated.

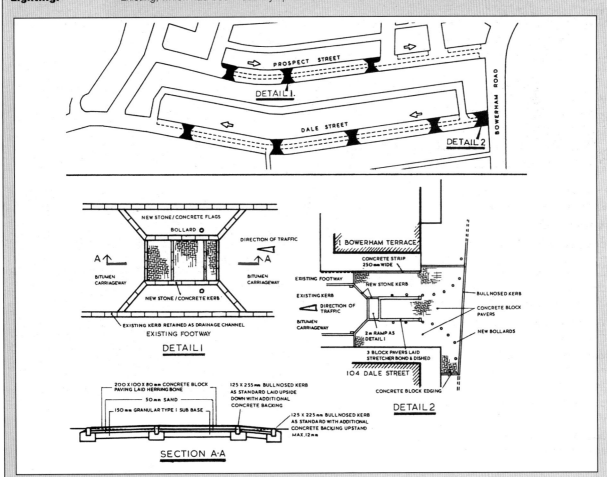

Costs: Lancashire County Council Small Improvements budget £30,600.
Lancashire County Council Highways Maintenance Allocation £8,400.
Total £39,000.

Contact's Comments: Effective speed reduction measures and redistribution of traffic in a more acceptable manner.

76 RAISED PINCH POINTS

RESIDENTIAL

Burnley, Brougham Street

Location: Lancashire: Burnley, Brougham Street.

Implemented: May 1992.

Background: Brougham Street is used as a rat-run for commuters gaining access to the M65.

Need for Measures: Child pedestrian accidents, primarily to resident ethnic population as a result of speeding.

Measures Installed: Two slightly raised single way working pinch points, approximately 30 m in length protected by a rumble area at one end and a chicane created by a refuge and build-outs.

Special Features: Rumble area and chicane.

Consultation: With bus companies, police, local residents (leaflets and public meetings) and Burnley Borough Council.

Technical Data:

Type of Location: Urban residential.

Road Type: Urban classified.

Type of Scheme: 2 single way working pinch points, slightly raised, with alternating give ways. 1 rumble area and 1 chicane as speed reducing features.

Total length of Scheme: 400 m.

Dimensions:

	Pinch points	Rumble area	Chicane
Height:	45 mm	45 mm	–
Width:	3.5 m	4.3 m	2.8 m
Length:	30 m	3 m	25 m

Materials: Pinch points: Marshalls brindle paving. Rumble Strips- Marshalls tegula blocks

Signs: Regulatory sign: Diags. 516 and 615. Informatory sign: Diag. 811. Road markings: Diags. 1003, 1023 and 1024

Lighting: Existing lighting upgraded.

Costs: Lancashire County Council Small Improvements budget £45,000. Lancashire County Council Accident Investigation and Prevention budget £2,000. Total £47,000.

Contact's Comments: None supplied.

Monitoring	Accidents (pia	Speeds (85% ile mph)	Traffic (24 hr av)
Before	19 in 3 years	35	5,592
After	5 in 1 yr 11 m	30	4,750

Contact: Ian Kime   Tel: 01772 264499

Authority: Lancashire County Council

77 VARIED MEASURES

RESIDENTIAL

Stockport: Woodsmoor

Location: Greater Manchester: Stockport, Woodsmoor, Moorland Road and Woodsmoor Lane.

Implementation date: June 1992.

Background: Woodsmoor Lane and Moorland Road are both local distributor roads, forming a link between two strategic routes. This link is some 700 metres in length, subject to a 30 mph speed limit and has adequate lighting. Both roads are well established commuter rat-runs through residential 'Living Priority' areas, which also provide access to two local schools.

Need for Measures: Excessive speeds, which have led to injury and damage accidents, including one fatality, and to deter use as a regular rat-run.

Measures In stalled: Speed tables, flat top road humps, pinch points and priority changes.

Special Features: Chicanes and pinch points created by the use of raised planters and landscaping, ensuring low vehicle speeds and an aesthetically pleasing scheme.

Consultation: With the local schools and residents, emergency services and the relevant bus company.

Technical Data:

Location Type: Urban residential area.

Road Type: Local distributor road: 30 mph.

Type of Scheme: Speed tables, flat top humps, pinch points and the change of junction priorities.

Total Length of Scheme: 700 m.

Technical Specification of measures:
Height: 100 mm. Ramp gradient: 1:10.
Road widths: Varying from 3 m to 5.5 m.

Materials: Speed tables: Brindle block paving to denote flat top plateau. Ramps and humps: Buff block paving. Bollards: Black cast iron. Tree grilles: Black cast iron. Planters: Country rustic brick.

Signs: Warning signs to Diags. 501, 517 and 602. 'Road hump' sign Diag. 557.

Road Markings: 'Give way' associated markings Diags. 1003 and 1023. Hazard warning marking Diag. 1004. Hazard warning triangle on ramps.

Costs: Highways £55,000. Leisure Services £6,000. Street Lighting £9,000. Total £70,000.

Contact's Comments: After initial local criticism the scheme has settled down fairly well.

Monitoring	Accidents (pia)	Speeds (mph)	Traffic
Before	23 in 3 years	38	2,875
After	1 to date	22	1,650

Contact: Colin Robinson   Tel 0161 474 4849
Authority: Stockport Metropolitan Borough Council

Metropolitan Borough of Stockport
Technical Services Division

78 OPEN ROAD CLOSURE
Newham, Various Roads

Location: London: Newham, various roads.

Implemented: 1981 onwards. This example 1987.

Background: The roads selected were all residential and suffering from high levels of extraneous traffic. They were chosen following complaints and petitions from residents for the roads to be closed.

Need for Measures: To remove unnecessary traffic without creating long diversions for cyclists, refuse collection and emergency vehicles.

Measures Installed: Road narrowed to single track, 1.2 m wide, for use by cyclists. The rest of the carriageway raised 150 mm using 1 m raised strips of deterrent paving either side of the cycleway and clay blockwork over the remainder. The 1 m raised strips are accessed by concrete ramps. Trees are planted on the footways either side.

Special Features: Deterrent paving. Blockwork. 'No entry' sign with exemption plate for emergency and refuse vehicles. Landscaping and trees.

Consultation: Local residents by questionnaire and 10% sample interviews. Emergency services.

Monitoring	Accidents	Speeds (mph)	Traffic (12 hrs)
Before	6 in 3 years	28	1,700
After	0	0	90

Contact: Ray Bradley   Tel: 0181 4762 1430 ext 22357
Authority: Newham Council

NEWHAM COUNCIL
THE HEART OF EAST LONDON

RESIDENTIAL

Technical Data:

Location Type: Residential area.

Road Type and Speed Limit: Urban unclassified: 30 mph.

Scheme Type: Road closure with exemptions.

Length of Scheme in Total: This street 500 m long.

Dimensions: Height: 150 mm.
Width: Cycleway 1.2 m. Ramps 1 m.
Length: Plateau 10 m. Ramps 1 m.
Gradient: 1:10, and 1:20 on bus routes.

Materials: Plateau: Precast concrete deterrent paving, and clay blocks.
Ramps: Concrete.
Kerbs: Precast concrete.

Signs: Diags. 616 and 619, illuminated.

Lighting: No new lighting

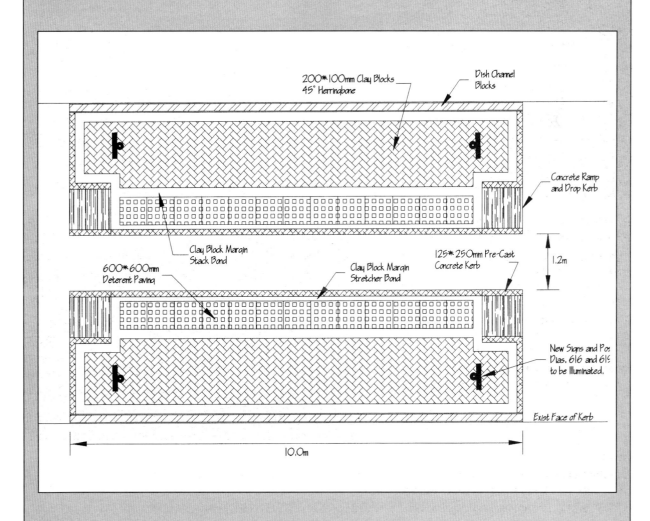

Cost: £10,000 per closure.

Contact's Comments: 95% compliance. Offenders almost all local residents. Hopper Bus now using
at 15 minute frequency after modification to make ramp gradients 1:20. No moving parts, such as gates, to maintain.
Some loss of on-street parking.

79 FLAT TOP HUMPS, ZEBRA
Banbury, The Fairway

Location: Oxfordshire: Banbury, The Fairway.

Implemented: June 1992.

Background: The Fairway is a distributor road in a large post-war residential area. It is approximately 0.8 km in length, subject to a 30 mph speed limit and has lighting. Prior to the completion of the M40 extension, congestion on the trunk roads in Banbury resulted in significant rat running traffic in The Fairway.

Need for Measures: Speed measurements showed a significant level of abuse of the 30 mph limited, even after the M40 extension was opened. A very high proportion of the casualties recorded in the road were child pedestrians.

Measures Installed: 11 flat top road humps including 1 humped zebra crossing.

Special Features: The road is used by a relatively high frequency bus service. Considerable concern was expressed by the operator, particularly as the same service has to traverse a road hump scheme in a nearby street which employed severe (100mm) flat top humps. In view of this, the emphasis was to provide road humps which were both effective and as bus friendly as possible.

Consultation: With police, emergency services, bus operator, District Council and public (through public display).

Monitoring	Accidents (pia)	Speeds (mph)	Traffic (24 hr)
Before	4 in 3 years	34	5,700
After	4 in 2 years	26	4,200 [1]

[1] Decrease in flow also attributed to M40 traffic relief

Contact: Ann Mortlock Tel: 01865 810400

Authority: Oxfordshire County Council

Technical Data:

Location Type: Residential.

Road Type and Speed Limit: Urban unclassified: 30 mph.

Scheme Type: Flat top road humps.

Length of Scheme in Total: 0.8 km.

Dimensions: Height: 60 mm.
Width: 6 m.
Length: Plateau 3.2 m. Ramps 720 mm.
Ramp gradient: 1:12.

Materials: Hot rolled asphalt.

Signs: Road hump and humped zebra crossing warning signs.

Lighting: No additional lighting.

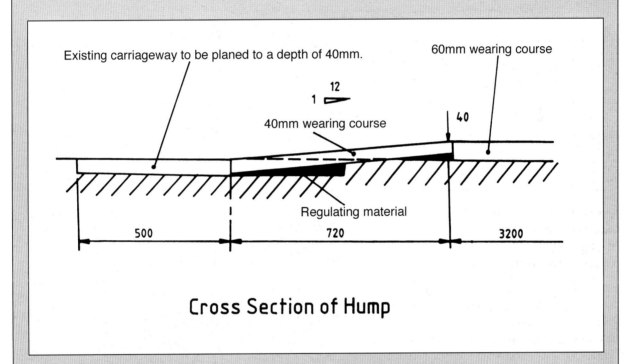

Existing carriageway to be planed to a depth of 40mm.

60mm wearing course

40mm wearing course

Regulating material

500   720   3200

Cross Section of Hump

Cost: £17,500 approximately.

Contact's Comments: Despite the apparent increase in accident frequency there have been no reported injury accidents involving child pedestrians and none of the recent accidents can be attributed to the scheme. In the 'before' period 3 of the 4 reported casualties were children. The accidents after the scheme was completed were of miscellaneous type and cannot be attributed to the scheme.

80 NARROWING, CHICANES, 'THUMPS'
Drumchapel, Peel Glen Road

Location: Strathclyde: Glasgow, Drumchapel, Peel Glen Road.

Implemented: May 1992.

Background: Drumchapel is a large estate on the outskirts of Glasgow, comprising post 1945 local authority housing. Peel Glen Road is a minor unclassified road forming part of a route between the A810 Duntocher Road and the A82 Great Western Road.

Need for Measures: High speeds, which led to child pedestrian accidents; and rat running. One survey revealed that 82% of the traffic on Peel Glen Road during the morning peak was city bound through traffic.

Measures Installed: Chicanes, speed throttle, thermoplastic road 'thumps', additional road markings, large advisory signs, warning signs, pedestrian barrier rail and reflective concrete bollards.

Special Features: The use of thermoplastic road 'thumps' as a cheaper alternative to conventional road humps and use of specific advisory signs. Both the ramps and the signs required to be authorised by the Scottish Office.

Consultation: Police and other emergency services, local Regional councillor, Tenants Association, Drumchapel Initiative and the Strathclyde Passenger Transport Executive.

Monitoring	Accidents (pia)	Speeds (mph [1])	Traffic (10 hr)	
			peak hour	10 hour
Before	8 in 3 years	30	700	2,765
After	1 in 2.5 years	18	400	1,895

1 Average of Northbound at one site and Southbound at two sites

Contact: Ken Aitken   Tel: 0141 227 2573

Authority: Strathclyde Regional Council

Strathclyde

Technical Data:

Location Type: Urban, peripheral housing estate.

Road Type and Speed Limit: Urban unclassified: Mainly 30mph with de-restricted section.

Scheme Type: Narrowings, chicanes - single way working, thermoplastic 'thumps'.

Length of Scheme in Total: 420 m.

Dimensions: Chicane: Length 23.7 m. Reduced carriageway width 5.5 m. Footway extensions 6 m radius
extending to centre of carriageway.

Throttle: Length 11.3 m. Reduced carriageway width 3.5 m. Footway extension width 2.1 m
(including 0.5 m channel).

Thermoplastic 'thumps': Stepped with base layer 200 mm x 10mm and top layer 100 mm x 20 mm.
Laid in series with spacing between ramps diminishing as physical alteration is approached.

Materials: Chicanes, gateways, islands: Asphalt with concrete kerbs and bollards.
Road ramps: White thermoplastic material.

Signs: Chicanes: Warning sign Diag. 516 with plate Diag. 518.
Throttle: Regulatory sign Diag. 615. Informatory sign Diag. 811.
Advisory 'Traffic Calming Scheme' sign - non prescribed

Road Markings: White lining at chicanes and throttle to make road appear narrower.

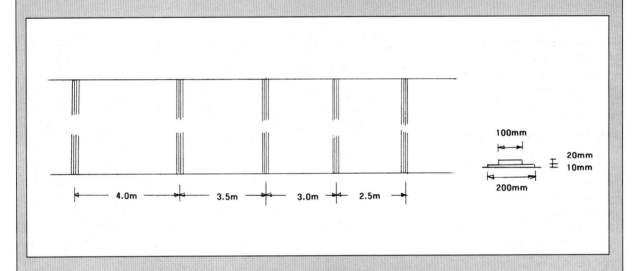

Lighting: As existing.

Cost: Strathclyde Roads Department Capital budget £15,000.

Contact's Comments: This scheme has been successful in achieving reduction of accidents, speeds and through traffic, and the use of road ramps has been extended to adjacent roads. The one pedestrian accident after implementation was not related in any manner to the traffic calming measures which were introduced in Peel Glen Road.

81 ROUND TOP HUMPS
Guildford, Cumberland Ave

Location: Surrey: Guildford, Cumberland Avenue.

Implemented: January 1991.

Background: Cumberland Avenue is a residential `D' road which provides an attractive cut-through for motorists trying to avoid delays at the junction of Salt Box Road and the A322.

Need for measures: Speed related accidents on a bus route inappropriate for through traffic.

Measures installed: 12 round top road humps along the full length of the road.

Special Features: Humps with tapered edges to assist drainage.

Consultation: Limited study initiated following request from residents association. (Surrey CC now undertakes thorough public consultation during the early stages of all traffic calming schemes).

Monitoring	Accidents (pia)	Speeds[1] (mph)	Traffic Flow (am pk hr)
Before	4 in 3 years	34	360
After	0 in 3 years	19	240

[1]Average freeflow speeds at humps and fastest point between are 18mph and 20mph.

Contact: Clive Batchelor Tel: 0181 541 9326

Authority: Surrey County Council

SURREY COUNTY COUNCIL

Technical Data:

Location Type: Residential.

Road Type and Speed Limit: Unclassified: 30 mph.

Scheme Type: Round top road humps.

Length of Scheme: 700 m approximately.

Dimensions: Round top humps: in accordance with Road Hump Regulations 1990.
Hump height: 75mm; Hump length: 3.7m.
Spacing of humps: varies between 36m and 83m.

Materials: Asphalt.

Signs: 'Road hump' sign.

Lighting: General lighting improvements required.

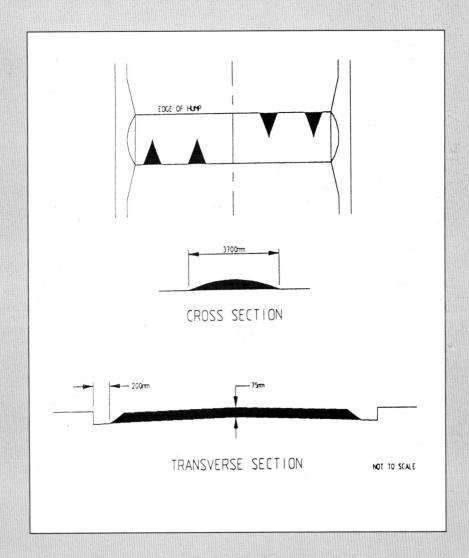

EDGE OF HUMP

3700mm

CROSS SECTION

200mm   75mm

TRANSVERSE SECTION   NOT TO SCALE

Cost: Funded by Surrey County Council £35,000 approximately.

Contacts Comments: Scheme successful in reducing accidents and speed. Through traffic was also reduced and the results of a public opinion survey suggest that most residents are pleased with scheme.

82 CHICANES
Wakefield, Queen Elizabeth Road

Location: West Yorkshire: Wakefield, Queen Elizabeth Road.

Implemented: September 1992.

Background: Queen Elizabeth Road is an unclassified distributor road from the A642 into the Eastmoor housing estate. Residential properties and a park/recreational area front the road which serves as a main access to a school. It varies in width between 4.9 and 6.5 metres and is lit. The road is used regularly by emergency services with the fire and ambulance stations being located nearby, and is also a bus route.

Need For Measures: To reduce vehicle speeds and carriageway widths in an attempt to address accidents, particularly to pedestrians and cyclists.

Measures Installed: A series of chicanes complemented with signing and lining and ultimately with upgraded street lighting.

Special Features: The use of reflective rubber kerbs and bollards bolted to the carriageway which allows implementation without the need for kerb foundation, excavation etc.

Consultations: Emergency services, bus operators, motoring organisations and residents.

Monitoring	Accidents (pia)	Speeds (85%ile mph)	Within limit (%)	Traffic (2 way)
Before	7 in 5 years	38	25	3,000
After	1 in 22 months	30	90	2,700

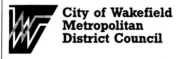

City of Wakefield Metropolitan District Council

Contacts: Mr A M Salmon   Tel: 01924 296066

Authority: Wakefield Metropolitan District Council

Technical Data:

Location Type: Residential area, school, emergency service and bus route.

Road Type and Speed Limit: Urban unclassified distributor: 30 mph.

Scheme Type: Reflectorised rubber kerbs and bollards used to provide a chicane effect.

Length of Scheme in Total: 500 m.

Dimensions: Height: Kerb 100 mm.
Width: 1.14 m.
Length: 2 m.

Materials: BTM (Safety Products) rubber kerbs and bollards. Infill with 75 mm crushed stone
and 25 mm bituminous macadam wearing course.

Signs: Priority signing to Diags. 811 and 615.
Carriageway markings to Diags 1003, 1004, 1023, 1038, and 1040.

Lighting: Upgraded from 5 m concrete LP sodium to 6 m steel 70 watt HP sodium.

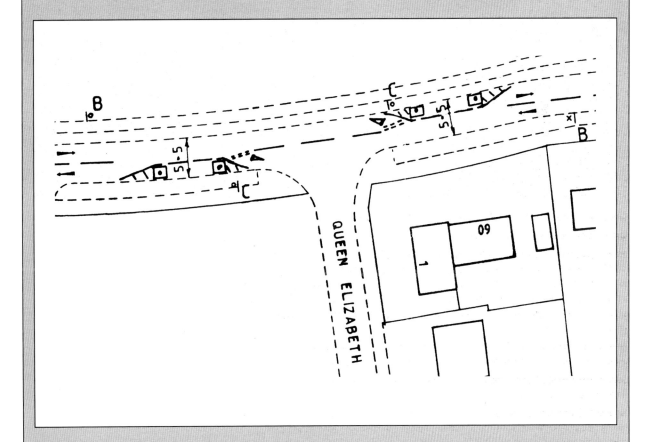

Cost: Chicanes £10,000.
Street Lighting. £15,000.
Total £25,000.

Contact's Comments: The scheme was devised because the emergency services opposed conventional vertical displacement calming measures such as road humps. Problems were experienced with inadequate conspicuity of the chicanes. To address this, upgraded street lighting has been installed and the rubber bollards have been replaced with low mounted chevron boards.

83 CHICANES, NARROWING, CYCLE TRACK, PROTECTED PARKING
Woking, Albert Drive

Location: Surrey: Woking, Albert Drive.

Implemented: August 1991.

Background: Albert Drive is a mainly residential 'D' road, approximately 2.4 km long, with industrial units at west end and a school of 535 pupils at the east end. It provides an attractive alternative to the A245, the designated route for Woking town centre.

Need for measures: Accident problem with excessive speed. A bus route inappropriate for through traffic.

Measures installed: Three chicanes as shown, with a cycletrack, at the east end. Ramped protected parking, chicane, and central hatching along the remainder of Albert Drive.

Special features: Additional kerb build out on exit of chicane to provide additional speed reduction. Chicane dimensions finalised by on-site bus trials. Chicane island designed as pedestrian crossing point.

Consultation: Emergency services. Press release. No formal public consultation carried out by Woking BC. (Surrey CC now undertakes thorough public consultation during the early stages of all traffic calming schemes).

Monitoring	Accidents (in Chicane section)	Speeds		Traffic (24 hr)
		>30mph	>40mph	
Before	36 (5) in 3 yrs	75%	16%	6,334
After	10 (1) in 2.5 yrs	23%[1]	1%[1]	5,761

[1] between chicanes. Average freeflow speeds at chicanes are 25 mph, and at fastest point between is 32 mph

SURREY COUNTY COUNCIL

Contact: Clive Batchelor **Tel:** 0181 541 9326

Authority: Surrey County Council

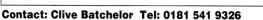

Technical Data:

Location Type: Residential.

Road Type and Speed Limit: Unclassified: 30 mph.

Scheme Type: Chicane system.
Special authorization for use of non-standard sign.

Length of scheme: 500 m approximately.

Dimensions: Carriageway width on entry to chicane: 5 m.
(3 m width available for each direction of flow at chicane mid point).
Length of chicane: 21.6 m.
Distance between chicanes: 150 m (average).

Materials: Standard construction materials.

Signs: Use of chicane sign on each approach as detailed above.

Lighting: General lighting improvements required.

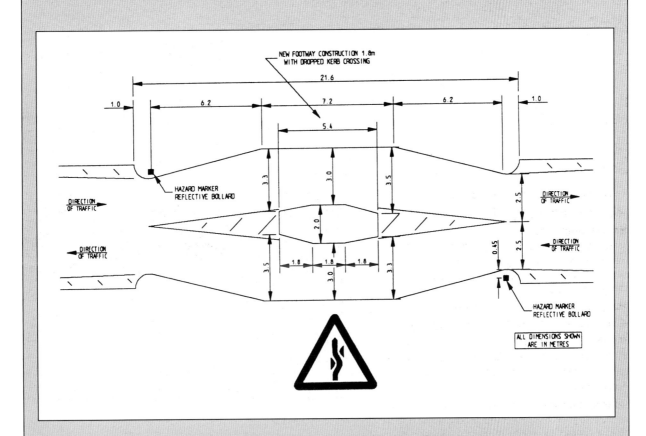

Cost: Jointly funded by Surrey County Council and Woking Borough Council.
Total £84,000 (including £17,900 for streetlighting improvements).

Contact's Comments: Successful in reducing accidents and speed. Cycletrack removes danger
to cyclists at chicanes. A public opinion survey showed that residents have mixed views. The bus friendly design
could be adapted elsewhere when a similar speed reduction is required.

84 THERMOPLASTIC 'THUMPS'
South Elmsall, Ash Grove

Location: West Yorkshire: Wakefield, South Elmsall, Ash Grove.

Implemented: February 1993.

Background: Ash Grove is an unclassified road with residential properties and shops on one side with a school, swimming baths and residential homes (some for the elderly) on the other. The road is lit, the average width is 7 metres and it is subject to a 30 mph speed limit. The road is also a bus route and a direct link between two main roads.

Need for Measures: To reduce vehicle speeds on a busy residential road to assist pedestrians and reduce injury accidents.

Measures Installed: Thermoplastic mini humps and associated signing.

Special Features: The use of thermoplastic 'thumps', which are designed to be smaller than conventional humps, thus minimising the effect on emergency services and bus operators whilst still achieving a reduction in vehicle speed. This was a trial scheme installed in conjunction with the DOT which funded the installation costs.

Consultations: All emergency services, bus operators, Department of Transport, motoring organisations and residents of fronting properties.

Monitoring	Accidents (pia)	Speeds (85%ile mph)	Traffic (vpd)
Before	5 in 5 years	33	3,000
After	1 in 1.5 years	25	2,600

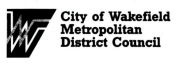

City of Wakefield
Metropolitan
District Council

Contact: Mr A M Salmon Tel: 01924 296066
Authority: Wakefield Metropolitan District Council

Technical Data:

Location Type: Residential, school, community uses.

Road Type and Speed Limit: Urban unclassified distributor: 30 mph.

Scheme Type: Thermoplastic 'thumps' and associated signing. Special authorisation obtained from the Department of Transport.

Length of Scheme in Total: 750 m.

Dimensions: Height: 37 mm.

Width: 7 m.

Length: 900 mm.

Gradient: Curved segment with a 2.68 m radius.

Materials: Thermoplastic material complying with BS 3262. Colour: Yellow complying to BS 381C No. 355. Marking has final film with Ballotini and conforms to the requirements of BS 3262 Clause 15.

Signs: Triangular road hump warning signs with associated distance plate and directional signing where suitable.

Lighting: Existing lighting system maintained.

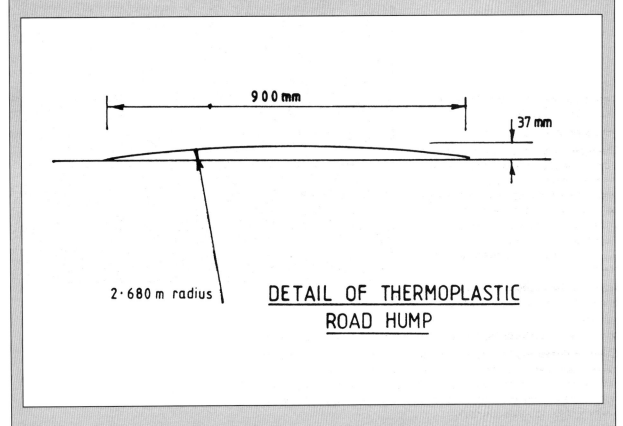

900 mm

37 mm

2·680 m radius

DETAIL OF THERMOPLASTIC
ROAD HUMP

Cost: £13,500, met by the Department of Transport on the basis that the scheme was a trial site.

Contact's Comments: The fire service and bus operators originally objected to the provision of conventional road humps hence the development of the smaller thermoplastic hump. Whilst the fire service has indicated that it is satisfied with the scheme, the bus operators threatened to remove services from Ash Grove due to increased vehicle maintenance costs. A survey of local residents also showed them to be in favour of removing the 'thumps' and they are to be removed in the near future and replaced by a series of chicanes.

85 SINGLE AND TWO-WAY RAMPS
Laverstock, Church Road

Location: Wiltshire: Laverstock, Church Road.

Implementation date: August 1992.

Background: Church Road is on the periphery of Salisbury. It is subject to a 30 mph speed limit and has lighting. It is used as a commuter cut-through to Salisbury centre and also to bypass Salisbury and link the A30 and A36.

Need for measures: There was a concentration of accidents in the vicinity of three secondary schools and a primary school. Also the scheme was to cater for a proposed cycle/pedestrian track to Church Road for the school children.

Measures installed: 2 single-way working ramps on entry to the system and 2 two-way ramps.

Special features: One of the two-way humps was designed to form a crossing point for the proposed cycle pedestrian track.

Consultation: Local schools, police, District and Parish councils, bus companies and local residents.

Monitoring	Accidents (pia)	Speeds (mph)	Traffic
Before	7 in 6 years	38.5	4,000
After	0 in 14 months	24.5	4,400

Contact: Tim Little   Tel: 01225 713 462

Authority: Wiltshire County Council

Technical data

193

Location Type: Urban fringe.

Road Type and Speed Limit: Urban unclassified: 30 mph.

Type of Scheme: Flat top humps with associated narrowings on entry to the scheme.
Cycle slips have been provided through the road narrowings.

Length of Scheme in Total: 410 m.

Dimensions: Height: 75 mm.
Width: 3 m and 6 m.
Length: Plateau 5.5 m.
Ramp gradient: 1:12.

Materials: Humps and kerbs: Recycled rubber.
Bollards: High intensity reflective.

Signs: Advance information signs, road humps signs,
road narrowing and priority signs, ramp warning signs.

Road markings: As per regulations.

Lighting: Additional column located at each hump.

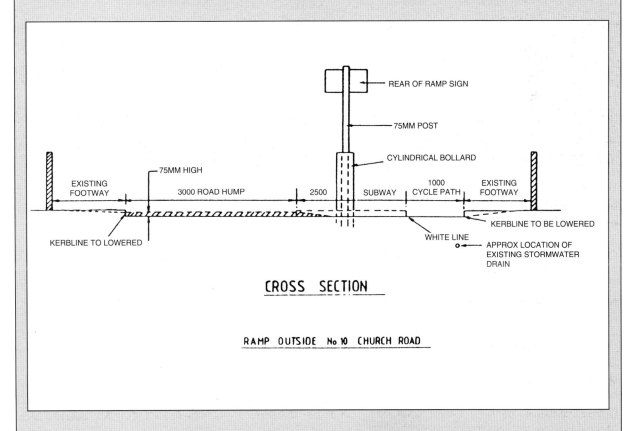

CROSS SECTION

RAMP OUTSIDE No 10 CHURCH ROAD

Costs: Wiltshire County Council Minor Improvements budget £22,000.

Contact's Coments: Performance of fixings for temporary rubber humps was unsatisfactory. Complaints about the noise created by lorries negotiating humps. Alternative measures under consideration.

GLOSSARY OF TERMS

A range of different terms is in use to describe the same types of traffic calming measures. This glossary briefly defines some common terms used in this book.

20 mph zone: An area wide zone of physical measures which reduces average speeds to below 20 mph. Subject to consent of the Secretary of State for Transport. Requires 18 month (minimum) experimental period of temporary consent to ensure measures have achieved the required speed reduction objective before permanent consent is granted.

Build out: A physical feature extending into the carriageway on one side only to narrow the road. If used to form a crossing for pedestrians, Section 75 of the Highways Act 1980 applies. Build outs can cause difficulties for cyclists.

Chicane: A form of *Horizontal Deflection*, using *Narrowing* formed by building out the kerbline, usually on alternate sides of the road to break up long straight stretches. The chicanes may be located so that, with priority signing, traffic is restricted to one way working. Two way working can also be retained in certain situations. Chicane layouts formed by *Build Outs* on one side only can also be devised either for one way or two way traffic depending on circumstances. By alternating parking areas from side to side of road a form of chicane can be produced, but only when cars are actually parked.

Cycle slip: Dedicated cycle lanes or 'bypasses' at *Pinch Points*, *Gateways*, *Chicanes* and where possible at *Road Humps*. They are important to ease the passage of cyclists.

Entry treatment: Treatment of side road junctions (bellmouths) to ensure drivers realise they are entering a road of a different character. May include *Gateways*, *Vertical Elements*, *Road Humps*, *Build Outs* and surface colour or texture changes.

Flat top road humps: The basic *Flat Top Road Hump* as prescribed in the Highway (Road Humps) Regulations 1990, will have tapered sides. Height 50mm minimum to 100mm maximum. Length of flat top 2.5m minimum. Length of ramp 600mm minimum. Ramp gradient 1:6 maximum; however 1:13 or flatter is becoming more common if humps have to be used on bus routes.

Gateway: Features provided, usually with *Vertical Elements*, to indicate to drivers where the character of a road changes. They can be at the start of a traffic calming scheme or on the entry to a village or rural settlement. Can be in combination with other measures such as *Rumble Devices*, *Pinch Points*, *Signs* and *Markings*.

Horizontal Deflection: General term for any measure that alters the horizontal alignment of the carriageway over a short distance .

Island: These are not pedestrian *Refuges* but physical islands without pedestrian facilities to assist in speed reduction by requiring vehicles to deflect from a straight path or by narrowing the carriageway. False roundabouts where no side road connections exist are one example. Islands can be incorporated at some *Gateway* sites. *Overrun Areas* can be incorporated.

Junction Platform: See *Raised Junctions*.

Mini Roundabout: Conventional mini roundabouts. Used at the start of, or within, a traffic calming scheme.

Narrowing: Restricting the width of available road space by physical measures such as *Build Outs*, *Pinch Points* and *Islands* or *Refuges*. Also by edge markings or centre hatching, colour or texture change to give drivers a strong visual impression of reduced width. A narrowing may be located so that, with priority signing, traffic is restricted to one way working.

Overrun Area: Area constructed by slightly raising the surface within the limits allowed by The Highway (Traffic Calming) Regulations 1993. Usually constructed in contrasting materials, to give the appearance of a narrower carriageway thereby inhibiting speeds. Larger vehicles can overrun these areas if necessary. Care in location and design is needed to deal with cycle and other two wheel traffic movements.

Pinch Point: A *Narrowing* formed by *Build Outs* opposite one another. Can be used in combination with *Gateways* and *Speed Cushions*. If intended to assist pedestrians to cross they are provided under Section 75 Of the Highways Act 1980.

Plateau: For the purpose of this publication, a type of *Flat Top Road Hump* with a flat top length of more than 6.0m, heights and ramps similar to *Speed Tables*.

Platform: Some authorities refer to extended *Flat Top Road Humps* as *Platforms* rather than as *Speed Tables* or *Plateaus*.

Protected Parking: Linear parking areas formed at the sides of roads by *Build Outs* which may be part of a *Chicane* system.

Raised Junction: The use of *Flat Top Road Humps* to raise whole junction areas. Also called 'Junction Platforms'. May be combined with a mini-roundabout but there is a requirement for a *Speed Reducing Measure* in advance of the raised area.

Refuge: As with *Islands* can be used to visually break up long straights, but provide for use by pedestrians and therefore come under Section 68 of the Highways Act 1980 not the Traffic Calming Act 1992.

Road Hump: Two basic types are recognised in The Highways (Road Humps) Regulations 1990 – *Round Top* and *Flat Top*. The latter are also described as *Speed Tables*, *Plateaus* and *Platforms*; they may be combined with *Horizontal Deflections* to permit single or two way traffic flow and may incorporate pedestrian crossings (Zebra or Pelican). *Raised Junctions*, *Platform Junctions* and raised *Entry Treatments* at junctions are all forms of *Flat Top Road Hump*. All road humps must be preceded by a *Speed Reducing Measure*.

Round Top Road Hump: Prescribed in The Highway (Road Humps) Regulations 1990 as with or without tapered sides (i.e. can be kerb to kerb); height 50mm minimum to 100mm maximum ; length 3.7m. Departures from this design requires DOT special authorisation.

Rumble Device: Measures that do not normally reduce traffic speeds in themselves, but which produce audible and vibratory effects to alert drivers to take greater care. Also called 'Rumble Strips', 'Rumble Areas' or 'Jiggle Bars' depending on layout and construction. Formed by a sequence of transverse strips laid across a carriageway, or areas of coarse surface dressing. Maximum permitted height of 15mm provided no vertical face exceeds 6mm. These measures are not normally acceptable in residential areas because of the associated noise.

Signs and Markings: For the purposes of this publication the terms refer to Signs and Markings used specifically with the aim of maintaining lower speeds, such as edge marking or centre hatching for carriageway *Narrowing*.

Speed Cushion: A *Round* or *Flat Top Road Hump* having a width less than the wheeltrack of a conventional bus but greater than an average car's wheeltrack. Can be used singly, in pairs or threes across the carriageway to suit circumstances. These devices require special authorisation by the DOT.

Speed Reducing Measure: These include physical features that drivers normally expect to encounter such as certain types of junction, roundabouts and specific degrees of bend which slow speeds down before drivers encounter a traffic calming measure. Traffic calming measures generally also reduce speeds but are for maintaining, not initiating, lower speeds.

Speed Table: For the purpose of this publication a type of *Flat Top Road Hump* with a flat top length of between 2.5m (the minimum) and 6.0m. Height 50mm minimum to 100mm maximum. A number of authorities have adopted a flat top length of 6.0m as 'standard' with a height of 75mm and ramp gradients between 1:10 and 1:15.

Thump: For the purpose of this publication these are mini-humps with a height greater than a *Rumble Device* but less than a *Road Hump*. Often made of a thermoplastic material. Require DOT special authorisation.

Vertical Deflection: General term for any measure that alters the vertical profile of the carriageway over a short distance. Maximum values are specified in The Highways (Road Humps) Regulations 1990.

Vertical Element: Vertical features such as signs, bollards, lighting columns, poles, trees that emphasise a change in road character at a *Gateway* or other traffic calming measure.

REFERENCES AND USEFUL READING

Statutes and Regulations

1. Highways Act 1980 and amendments. HMSO London

2. Highways (Traffic Calming) Regulations 1993. Statutory Instrument 1993 No. 1849. HMSO, London

3. Traffic Signs Regulations and General Directions 1994. HMSO, London

4. The Highway (Road Humps) Regulations. Statutory Instrument 1990 No. 703. HMSO, London

5. Roads (Scotland) Act 1994

6. The Road Humps (Scotland) Regulations 1990

Department of Transport Circular Roads

7. 3/90 Road Humps (Welsh Office 54/90)

8. 4/90 20mph Speed Limit Zones (Welsh Office 2/91)

9. 2/92 Road Traffic Act 1991: Road Humps and Variable Speed Limits (Welsh Office 46/92)

10. 1/93 Road Traffic Regulations Act 1984: Sections 81 - 85 Local Speed Limits (Welsh Office 1/93)

11. 2/93 The (Highways) Traffic Calming Regulations (Welsh Office 46/93)

Scottish Office Circulars

12. SOID Circular No 2/91. The Road Humps (Scotland) Regulations

13. SOID Circular No 7/92. The Use of Automatic Devices for the Detection of Speeding and Traffic Light Offences

14. SOID Circular No 8/92. Variable Speed Limits and Road Humps

15. SOID Circular No 10/92. 20mph Speed Limit Zones

16. SOID Circular No. 1/93.1990 Road Traffic Regulation Act 1984, Sections 81-85. Speed Limits

Traffic Advisory Leaflets (DOT Traffic Advisory Unit)

17. 1/87 Measures to Control Traffic for the Benefit of Residents, Pedestrians and Cyclists

18. 2/90 Speed Control Humps

19. 3/90 Urban Safety Management - Guidelines from IHT

20. 3/91 Speed Control Humps (Scottish Version)

21. 7/91 20mph Speed Limit Zones

22. 2/92 Carfax, Horsham - 20 mph Zone

23. 2/93 20mph Speed Limit Zone Signs

24. 3/93 Traffic Calming - Special Authorisations

25. 6/93 Traffic Calming Bibliography (superseded by TAL 5/94)

26. 7/93 Traffic Calming Regulations

27. 11/93 Rumble Devices

28. 12/93 Overrun Areas

29. 13/93 Gateways

30. 1/94 VISP A Summary

31. 2/94 Entry Treatments

(In Preparation)

3/94 Emergency Services and Traffic Calming: A Code of Practice

4/94 Speed Cushions

5/94 Traffic Calming Bibliography

7/94 'Thumps' – Thermoplastic Road Humps

9/94 Horizontal Deflections

Other References

32. A Case Study of British Traffic Calming Measures. Hass-Klau C (1990). *PTRC Summer Annual Meeting, Seminar B,* Bath, September·

33. *Accident Reduction and Prevention.* Institution of Highways & Transportation (1990) International Edition. IHT, London

34. Ambulances: Health Service Circular HSC(IS)67 (WHSC(SC)57) 1974. Department of Health (contains Standards of Performance). London

35. *An assessment of rumble strips and rumble areas.* Webster DC, Layfield RE (1993). Project Report 33, Transport Research Laboratory. Crowthorne, Berkshire

36. An experiment with road humps. Broadbent KA and Salmon AM (1991). *The Journal of Highway Incorporated Engineers,* November

37. *An Illustrated Guide to Traffic Calming.* Hass-Klau C (1990). Friends of the Earth. London

38. *An Improved Traffic Environment - A Catalogue of Ideas.* Road Directorate, Denmark Ministry of Transport (1993). Road Data Laboratory, Road Standard Division Report 106. Copenhagen

39. Appropriate Speeds for Different Roads and Conditions. Kimber RM (1990). *Conference on Speed Accidents and Injury: Reducing the Risks,* July. Parliamentary Advisory Council for Transport Safety, London

40. Burnthouse Lane Traffic Calming Scheme. Chorlton E (1990). *Highways and Transportation* 37(8), pp7-11

41. *Children and Roads: A Safer Way.* Department of Transport (1990). London

42. *Civilised Streets: A Guide to Traffic Calming.* Hass-Klau C, Nold I, Bocker G, Crampton G (1992). Environmental & Transport Planning, Brighton

43. *Code of Practice for Local Authorities/Emergency Services in the Preparation of Traffic Management Schemes.* Joint Traffic Executive and Association of London Borough Engineers & Surveyors (1983). Kingston upon Thames

44. *Cyclists and Traffic Calming.* Cyclists Touring Club (1991), CTC Technical Note. Godalming, Surrey

45. *Design Bulletin 32: Residential Roads and Footpaths - Layout Considerations.* Department of the Environment (1992) Second edition. DoE, London

46. *DOT Bypass Demonstration Project – Interim Reports.* Department of Transport, November 1992 and November 1993. London

47. *Dutch 30kph Zone Design Manual* (translation). Lines CJ and Castelijn HA (1991). Transport and Road Research Laboratory. TRL, Department of Transport

48. Environmental Traffic Management in Britain, Does it Exist? Hass-Klau C (1990). *Built Environment* 12(1/2), pp13-59

49. Environmental Traffic Management in Residential Areas - Experience with 30kph Zones in the Federal Republic of Germany. Schleicher-Jester F (1990). *International Congress on Living and Moving in Cities.* ADTS.CETUR, Paris

50. *Experimental Traffic Calming Scheme - Speed Cushions and Chicanes.* York City Council (1991). Unpublished

51. Fire Service: Home Office. Fire Service Circulars 43/1958 and 4/1985 (contain Standards of Cover for different areas of risk.)

52. *Guidelines for Urban Safety Management.* Institution of Highways and Transportation (1990). IHT, London

53. *Highways Economic Note No.1.* Department of Transport (1990). DOT, London

54. *Integrated Traffic Safety Management in Urban Areas.* OECD (1990). OECD, Paris.

55. Progress of Traffic Calming in German Towns and Cities. Keller H. (1991). *Transportation Planning Systems* 1(2), pp61-69.

56. *Public Attitude Survey – New Forest Traffic Calming Programme.* Windle R, Hodge A (1993). Transport Research Laboratory, Department of Transport

57. *Reducing Mobility Handicaps: Towards a Barrier-Free Environment.* Institution of Highways and Transportation (1991). Revised guidelines. IHT, London.

58. *Resume of Traffic Calming on Main Roads through Villages.* Wheeler AH (1992). Transport Research Laboratory, Department of Transport

59. *Roads and Traffic in Urban Areas.* Institution of Highways and Transportation (1987). HMSO, London

60. *Road Humps for Controlling Vehicle Speeds.* Webster DC (1993). Project Report 18, Transport Research Laboratory, Department of Transport.

61. *Road Humps Trials. A Joint Study.* Strathclyde Roads and Strathclyde Public Transport Executive (1993). Glasgow, Scotland

62. Safety Effects of Speed Reducing Measures in Danish Areas. Engel U (1990). *Seminar on Speed Management in Urban Areas.* Copenhagen

63. *Speed Control Humps - A Trial at TRL.* Hodge AR (1993). Project Report 32, Transport Research Laboratory, Department of Transport

64. *Speed Reduction in 24 Villages: Details from the VISP Study.* Wheeler A, Barker J, Taylor M (1994). Project Report 85, Transport Research Laboratory, Department of Transport

65. *Survey of Public Attitude Acceptability of Traffic Calming Schemes.* Windle R, Mackie A (1992). Transport Research Laboratory, Department of Transport

66. *The Effectiveness of Village Gateways in Devon and Gloucestershire.* Wheeler A, Taylor M, Payne A (1993). Project Report 35, Transport Research Laboratory, Department of Transport

67. *The FOE Guide to Traffic Calming in Residential Areas.* Friends of the Earth (1987). FOE, London

68. The UK Urban Safety Project - Design and Effect of Schemes. Mackie AM, Ward HA and Walker RT (1988). *PTRC Summer Annual Meeting, Seminar B,* Bath

69. *This Common Inheritance. Environment White Paper.* Department of the Environment (1990). HMSO, London

70. *Traffic Calming No.1 (1990-1992).* Current Topics in Transport: A Selection of Abstracts. Transport Research Laboratory, Department of Transport

71. *Traffic Calming Update No.1.1 (1992-1994).* Current Topics in Transport: A Selection of Abstracts. Transport Research Laboratory, Department of Transport

72. *Traffic Calming and Buses – Guidelines from London Transport.* Bus Priority and Traffic Unit, London Transport Buses (October 1993 and August 1994). LTB, London

73. *Traffic Calming Guidelines.* Devon County Council (1991). Exeter, Devon

74. *Traffic Calming.* County Surveyors' Society (1990). Report No. 1/10. Wiltshire County Council, Trowbridge, Wiltshire

75. Traffic Calming: An East London Case Study. Pharoah TM (1990). *Seminar on Speed Management,* Copenhagen

76. *Traffic Calming: A Code of Practice.* Kent County Council (1994) Third edition. KCC, Maidstone, Kent

77. *Traffic Calming: A New Framework for Urban Planning.* Pharoah TM (1990). *Conference on Traffic Calming,* Ealing, London.

78. Traffic Calming: The Future. Gercans R (1991). *PTRC Traffex Paper,* April. PTRC, London

79. Traffic Management - The Effect of Speed Cameras in West London. Swali LN, Department of Transport (1993). *PTRC Summer Annual Meeting,* Manchester

80. *Urban Safety Project, 1. Design and Implementation of Schemes.* Lynam DA, Mackie AM, and Davies CH (1988). Research Report 153, Transport & Road Research Laboratory, Department of Transport

81. *Urban Safety Project, 2. Interim Results for Area-wide Schemes.* Mackie AM and Ward HA (1988). Research Report 154, Transport and Road Research Laboratory, Department of Transport

82. *Urban Safety Project, 3. Overall Evaluation of Area-wide Schemes.* Mackie AM, Ward HA and Walker RT (1990). Research Report 263, Transport and Road Research Laboratory, Department of Transport

83. *Village Speed Control Working Group (VISP) - Final Report .* County Surveyors' Society, Department of Transport, Scottish Office, Welsh Office and Transport Research Laboratory (1994). DOT, London

84. SBS and SOBETMA: Environmental Traffic Management and the South Birmingham Study. Huddart KW, Wenban-Smith A and Pharoah T (1993). *PTRC Summer Annual Meeting, Seminar C,* Manchester